Beginner's Guide to SolidWorks 2010

Alejandro Reyes, MSME
Certified SolidWorks Professional

ISBN: 978-1-58503-573-1

PUBLICATIONS

Schroff Development Corporation

www.SDCpublications.com

Schroff Development Corporation

P.O. Box 1334

Mission KS 66222

(913) 262-2664

www.SDCpublications.com

Publisher: Stephen Schroff

Trademarks and Disclaimer

SolidWorks and its family of products are registered trademarks of Dassault Systemes. Microsoft Windows and its family products are registered trademarks of the Microsoft Corporation.

Every effort has been made to provide an accurate text. The author and the manufacturers shall not be held liable for any parts developed with this book or held responsible for any inaccuracies or errors that appear in the book.

Copyright © 2010 by Alejandro Reyes

All rights reserved. This document may not be copied, photocopied, reproduced, transmitted, or translated in any form or for any purpose without the express written consent of the publisher, Schroff Development Corporation.

IT IS A VIOLATION OF UNITED STATES COYRIGHT LAWS TO MAKE COPIES IN ANY FORM OR MEDIA OF THE CONTENTS OF THIS BOOK FOR EITHER COMMERCIAL OR EDUCATIONAL PURPOSES WITHOUT EXPRESS WRITTEN PERMISSION.

Examination Copies:

Books received as examination copies are for review purposes only and may not be made available for student use. Resale of examination copies is prohibited.

Electronic Files:

Any electronic files associated with this book are licensed to the original user only. These files may not be transferred to any other party.

Acknowledgements

The sixth revision to the *Beginner's Guide to SolidWorks* is dedicated to my lovely wife Patricia and my kids Liz, Ale and Hector, all of whom have always been very supportive, patient and comprehensive during the extensive writing of this book. To you, all my love.

Also wish to thank the hundreds of students, users, professors and engineers whose great ideas and words of encouragement have helped me improve this book and make it a great success.

About the Author

Alejandro Reyes holds a BSME from the Instituto Tecnológico de Ciudad Juárez, Mexico in electro-mechanical engineering and a Masters Degree from the University of Texas at El Paso in mechanical design, with strong focus in Materials Science and Finite Element Analysis.

Alejandro spent more than 8 years as a SolidWorks Value Added Reseller. During this time he was a Certified SolidWorks Instructor and Support Technician, CosmosWorks Support Technician, and a Certified SolidWorks Professional, credential that he still maintains. Alejandro has over 16 years of experience using CAD/CAM/FEA software and is currently the President of MechaniCAD Inc.

His professional interests include finding alternatives and improvements to existing products, FEA analysis and new technologies. On a personal level, he enjoys spending time with his family and friends.

Table of Contents

List of commands introduced in each chapter. Note that many commands are used extensively in following chapters after been presented.

PART MODELING

Housing:
New Part
Create Sketch
Confirmation Corner
Sketch Grid
Sketch Rectangle
Sketch Centerline
Sketch Relations
Smart Dimension
Sketch Status
Extrude Boss/Base
View Orientation
Mouse Gestures
Center Rectangle
Fillet
Magnifying Glass
Extruded Cut
Through All (End condition)
Sketch Fillet
Rename Features
Circle
Instant 3D
Mirror Features
Model Display Styles
Fly-Out Feature Manager
Dimension Tolerance
Hole Wizard
Cosmetic Threads
Sketch Point
Edit Sketch
Rebuild
Circular Pattern
Automatic Relations
Temporary Axes
Sketch Slot
Linear Pattern
Edit Material
Mass Properties

Side Cover
Revolved Boss/Base
Trim Entities
Extend Entities
Construction Geometry

Top Cover
Offset Entities
Mirror Entities (Sketch)
Up to Surface (End condition)
Shell
Measure Tool
Select Other
More Fillet options

Offset Shaft
Revolved Cut
Auxiliary Planes
Hide/Show sketch
Convert Entities
Polygon (Sketch)
Flip Side to Cut
Axis (Reference geometry)
Coordinate Systems

Worm Gear
Mid Plane (End Condition)
Chamfer
Dimension to Arc
Direction 2 (End Condition)

Worm Gear Shaft
General Review of commands

SWEEP AND LOFT

Sweep
Thin Feature
Auxiliary Plane at point
Sweep
Up to Next (End condition)
Full Round Fillet
Helix
Variable pitch helix
Guide Curves

Loft
Offset Plane
Hide/Show Plane
Loft
Start/End Conditions
Model Section View

DETAIL DRAWING

Housing Drawing
Part Configurations
Suppress Feature
Parent Child Relation
Unsuppress
Configure Dimensions
Change Configuration
New Drawing
Make Drawing from Part
View Palette (Drawing)
Projected View
Tangent Edge Display
Section View
Detail View
Model Items
(Import dimensions)
Drawing cleanup
Center Mark/Centerlines
Add Sheet to Drawing
Model View
    Configuration
Rename Sheet

Introduction

This book is intended to help new users to learn the basic concepts of SolidWorks and good solid modeling techniques in an easy to follow guide. It will be a great starting point for those new to SolidWorks or as a teaching aid in classroom training to become familiar with the software's interface, basic commands and strategies as the user completes a series of models while learning different ways to accomplish a particular task. At the end of this book, the user will have a fairly good understanding of the SolidWorks interface and the most commonly used commands for part modeling, assembly and detailing after completing a series of components and their 2D drawings complete with Bill of Materials. The book is focused on the processes to complete the modeling of a part, instead of focusing on individual software commands or operations, which are generally simple enough to learn. We strived hard to include the commands required in the **Certified SolidWorks Associate** test as listed in the SolidWorks website, and some more.

SolidWorks is an easy to use CAD software that includes many time saving tools that will enable new and experienced users to complete design tasks faster than before. Most commands covered in this book have advanced options, which may not be covered in this book. This is meant to be a starting point to help new users to learn the basic and most frequently used commands.

SolidWorks is one of the leading 3D mainstream design CAD packages, with hundreds of thousands of users in the industry ranging from one man shops to Fortune 500 companies, and with a strong presence in the educational market in high schools, vocational schools, and many of the leading universities in the world.

We'd love to hear from you, your experience with this book and any comments or ideas on how we can make it better for you by sending an email to: areyes@mechanicad.com.

Prerequisites

This book was written assuming the reader has knowledge of the following topics:

- The reader is familiar with the Windows operating system.
- Knowledge of mechanical design and drafting.
- Any experience with CAD systems is a plus.
- Understanding of mechanics of materials is a must for SimulationXpress.

Notes:

The SolidWorks Interface

The SolidWorks interface is simple and easy to navigate. The main areas in the interface include toolbars, menus, graphics area, Feature Manager/Property Manager and Task Pane. The 2010 version includes an intelligent system of pop up toolbars automatically activated when the user selects elements in the Feature Manager or graphics area. SolidWorks has icons and menus similar to those of Microsoft Office applications, and follows Windows rules like drag and drop, copy/paste, etc. The Menu bar automatically hides and is reveled again when the user moves the pointer over the SolidWorks banner in the upper left corner of the window, but can be made always visible by pressing the pin icon at its end as indicated.

Feature Manager **Menus and Toolbars** **View Toolbar** **Graphics Area**

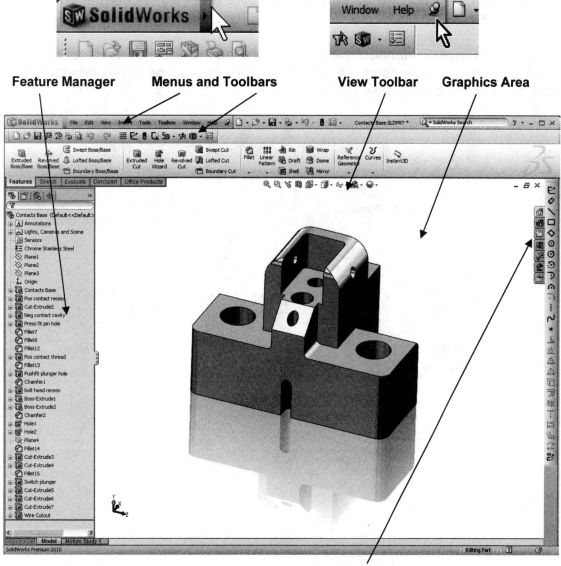

Task Pane Icons

The graphics area is the main part of SolidWorks and where most of the action happens; this is where parts, assemblies and drawings are created, visualized and modified. SolidWorks lets users Zoom, Pan, Rotate, change view orientations, etc., as well as to change how the models are displayed, either as Shaded, Hidden Lines, Hidden Lines Visible and Wireframe to name a few. The **Feature Manager** is the graphical browser of features and operations, where features can be edited, modified, deleted, etc. and is located in the left side of the screen.

The Feature Manager's space is also shared by the **Property Manager** and the Configuration Manager. This is where most of the SolidWorks' command options are presented to the user; this is also where a selected entity's properties are displayed and Configurations are created. The Property Manager is displayed automatically when needed, so the user does not need to worry about it. We'll show the user how to view both Feature and Property Managers at the same time later in the book.

In the Property Manager we have a common interface for most commands in SolidWorks, including common Windows Controls such as check boxes, open and closed option boxes, action buttons, etc.

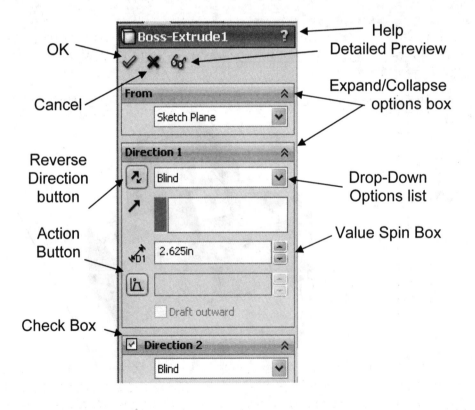

To manipulate the models in the graphics area, a set of tools is available from the menu, "**View, Modify**" to Zoom, Pan, Rotate, etc. Select a view manipulation tool, left click and drag the mouse in the graphics area to see its effect. The view can also be manipulated using the "View Orientation" drop-down toolbar at the top of the graphics area...

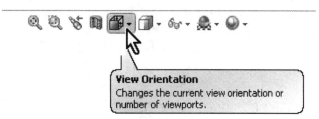

… Or the new **Mouse Gestures** introduced in SolidWorks 2010. Gestures are activated by right-dragging in the graphics area, the gestures shortcuts wheel will appear, and we simply keep dragging to touch the command we want. To modify the commands in the wheel we have to select the menu "**Tools, Customize**" and select the Mouse Gestures tab. By default mouse gestures are enabled, but they can be turned off in this tab; we can configure either 4 or 8 gestures for each environment (Part, Assembly, Drawing and Sketch) by selecting the command for each environment from the list of commands and selecting the gesture to assign. In this book we'll use the 8 gesture default settings.

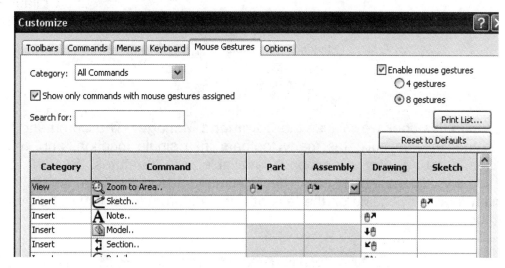

The Mouse scroll wheel can be used to zoom in and out in the model (the model is zoomed in where the mouse pointer is at), and clicking the Middle Mouse Button (the wheel) and dragging the mouse **Rotates** the part in the graphics area. The rotation is automatic about the area of the model where the pointer is located. The **Previous View** and **Zoom to fit** (default shortcut "F") are single click commands in the view orientation toolbar.

Use the **Standard Views** icon to view the model from any orthogonal view (Front, Back, Left, Bottom, Top, Right, Isometric, etc). Another way to rotate the models on the screen is with the arrow keys in the keyboard. Holding down the "Shift" key with the arrow keys rotates the model in 90° increments.

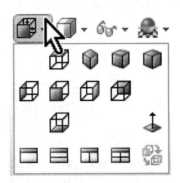

To change the **Display Style** (the way the model looks in the graphics area), the Display Style icon can be selected in the View toolbar; the effects will be immediately visible to the user. Feel free to explore them with your first model to become familiar; sometimes it's convenient to switch to a different view style for visibility or easy selection of internal or hidden entities.

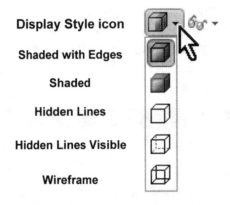

In this book we will use the **Command Manager**. The Command Manager is a tool that consolidates many toolbars in a single location, and selecting a toolbar's tab displays commands available, like Features, Sketch, Detailing, Assembly, Sheet Metal, etc. The Command Manager is a smart feature in SolidWorks; depending on the task at hand, different toolbars will be available to the user.

To enable the Command Manager select the menu **"View, Toolbars, Command Manger"**. You must have a document open to be able to turn the Command Manager on or off. For clarity, the option "Use Large Buttons with Text" has been enabled; it can be activated by Right-Mouse-Clicking anywhere in the Command Manager and selecting the option from the pop up menu.

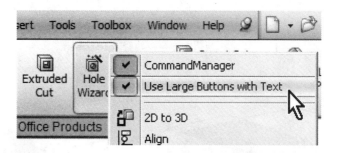

Models in SolidWorks can be displayed as simple solid colors or with high quality images; depending on the video card (graphics accelerator) used, real time reflections and shadows can be displayed using "Real View" technology. For clarity purposes Real View images will be used only from time to time in this book to help the reader more easily understand the concepts presented.

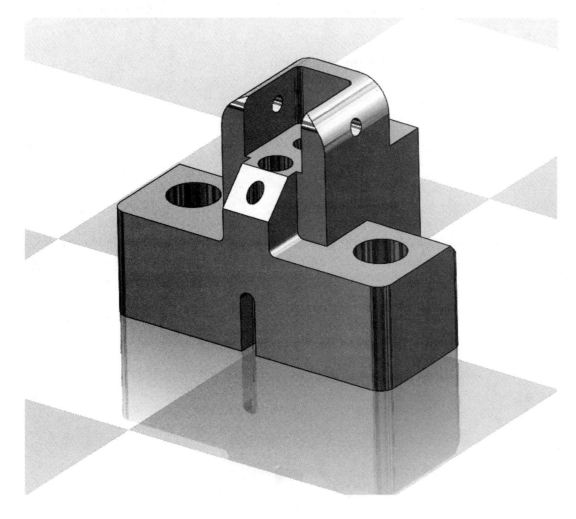

One option the user may wish to set is to view dimensions flat to screen, this way, regardless of the orientation of the part, the dimensions will be easier to read. This option can be found in the menu "**Tools, Options**" under the "System Options" tab, in the "Display/Selection" section. This book will use the "Display dimensions flat to screen" option on.

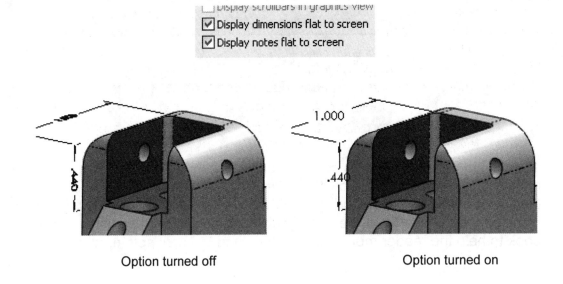

| Option turned off | Option turned on |

With that said, let's design something…

Part Modeling

The design process in SolidWorks generally starts in the part modeling environment, where we create the different parts that make the design of the product or machine, and are later assembled to other parts; at that time the group of parts becomes an Assembly. In SolidWorks, every component of the design will be modeled separately, and each one is a single file with the extension *.sldprt. SolidWorks is a *Feature based software*; this means that the parts are created by incrementally adding features to the model. Features are operations that either add or remove material to a part, for example, extrusions, cuts, rounds, etc. There are also features that do not create geometry, but are used as a construction aid, such as auxiliary planes, axes, etc.

This book will cover many different features to create parts, including the most commonly used tools and options. Some features require a **Sketch** or profile to be created first; these are known as Sketched features. The Sketch is a 2D environment where the sketch or profile to generate a feature is created. It is in the Sketch where most of the design information is added to the design, including dimensions and geometric relations. Examples of sketched features include Extrusions, Revolved features, Sweeps and Lofts.

A 2D Sketch can be created only in Planes or planar (flat) faces. By default, every SolidWorks Part and Assembly has three **default planes** (Front, Top and Right) and an Origin. Most parts can be started in any one of these planes. It is not critical which plane we start our designs in; however, the plane's initial selection can potentially save us a little time when working in an assembly or while detailing the part in the detail drawing for manufacturing. For a few releases of SolidWorks now, the plane selection is a lot less significant, as the detailing environment is easier to use and understand. The planning that takes place before starting to model a part is called the ***Design Intent.*** The Design Intent basically includes the general plan of how the part is going to be modeled, and how we anticipate it may change to accommodate future changes to fit other parts in an assembly or needs.

SolidWorks is a 3D parametric mechanical design software. **Parametric** design means that the models created are driven by parameters. These parameters are dimensions, geometric relations, equations, etc. When any parameter is modified, the 3D model is updated and the new parameters are reflected. Good design practices are reflected in how well the Design Intent and model integrity is maintained when parameters are modified. In other words, *the model updates predictably.*

Notes:

The Housing

When we start a new design, we have to decide how we are going to model it. Remember that the parts will be made one feature or operation at a time. It takes a little practice to define the optimum feature sequence for any given part, but this is something that you will master once you learn to think of parts as a sequence of features or operations. To help you understand how to make the Housing part, we'll show a "roadmap" or sequence of features. The order of some of these features can be changed, but remember that we need to make some features before others. For example, we cannot round a corner if there are no corners to round! A sequence will be shown at the beginning of each part, and the dimensional details will be given as we progress.

In this lesson we will cover the following tools and features: creating various sketch elements, geometric relations and dimensions, Extrusions, Cuts, Fillets, Mirror Features, Hole Wizard, Linear and Circular Patterns. For the Housing, we'll follow the following sequence of features:

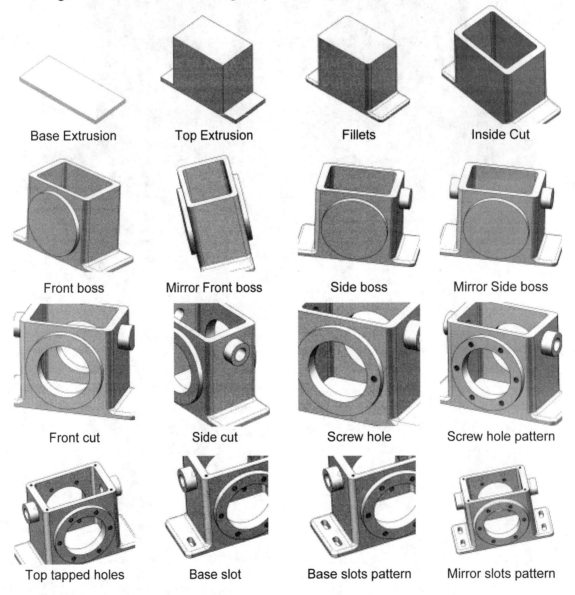

Base Extrusion	Top Extrusion	Fillets	Inside Cut
Front boss	Mirror Front boss	Side boss	Mirror Side boss
Front cut	Side cut	Screw hole	Screw hole pattern
Top tapped holes	Base slot	Base slots pattern	Mirror slots pattern

1. -   The first thing we need to do after opening SolidWorks, is to make a **New Part** file. Go to the "New" document icon in the main toolbar and select it.

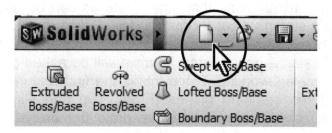

2. -   We are now presented with the New Document dialog. If your screen is different than this, click the "Novice" button in the lower left corner. Now select the "Part" template, and click OK, this is where we tell SolidWorks that we want to create a Part file. Additional Part templates can be created, with different options and settings, including different units, dimensioning standards, materials, colors, etc. See the Appendix for information on how to make additional **templates** and change the document **units** to inches and/or millimeters. Using the "Advanced" option allows the user to choose from different custom templates when creating new documents.

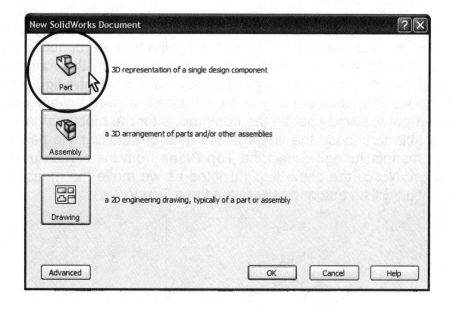

3. - Now that we have a new Part file, we have to start modeling the part, and the first thing we need to do is to make the extrusion for the base of the Housing. The first feature is usually one that other features can be added to or one that can be used as a starting point for our model. Select the "Extruded Boss/Base" icon from the Command Manager's Features tab (active by default). SolidWorks will automatically start a new **Sketch,** and we will be asked to select the plane in which we want to start working. Since this is the first feature of the part, we will be shown the three standard planes (Front, Top and Right). Remember the sketch is the 2D environment where we draw the profile before creating an extrusion, in other words, before we make it "3D".

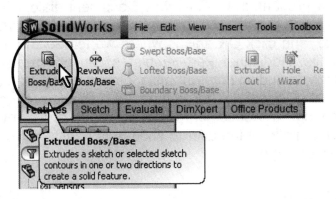

4. - For the Housing we'll select the "Top Plane" to create the first sketch. We want to select the Top Plane because we are going to start modeling the part at the base of the Housing and build it up as was shown in the roadmap at the beginning of this chapter. Do not get too worried if you can't figure out which plane to choose first when starting to model a part; at worst, what you thought would be a Front view may not be the front; this is for the most part irrelevant, as the user is able to choose the views at the time of detailing the part in the 2D drawing for manufacturing. Select the **Top Plane** from the screen using the left mouse button. Notice the plane is highlighted as we move the mouse to it. The view orientation will be automatically rotated to a Top View.

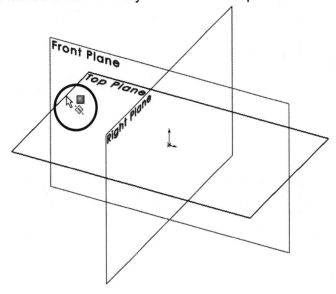

What we have just done is we created a new **Sketch** and are now in the sketch environment. This is where we will create the profiles that will be used to make Extrusions, Cuts, etc. SolidWorks gives us many indications, most of them graphical, to help us know when we are working in the Sketch environment.

a) The **Confirmation Corner** is activated in the upper right corner and displays the Sketch icon in transparent colors.

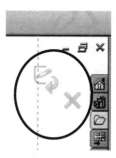

b) The Status bar at the bottom shows "Editing Sketch" in the lower right corner.

c) In the Feature Manager "Sketch1" is added at the bottom just under "Origin".

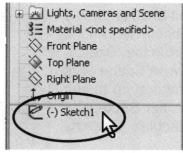

d) The part's Origin is projected in red and the grid is visible.

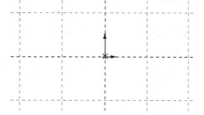

e) The Sketch tab is activated in the Command Manager displaying sketch tools.

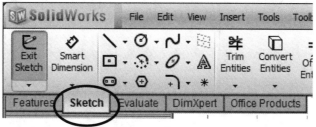

f) If the option is activated, the Sketch Grid will be displayed. This can be easily turned on or off *while in the Sketch environment*, by Right-Mouse-Clicking in the graphics area and selecting the "Display Grid" option.

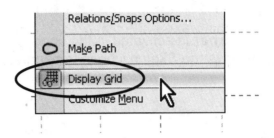

As the reader can see, SolidWorks gives us plenty of clues to help us know when we are working in a sketch.

5. - Notice that when we make the first sketch, SolidWorks rotates the view to match the plane that we selected. In this case, we are looking at the part from above. <u>This is done only in the first sketch</u> to help the user get oriented. In subsequent operations we have to rotate the view manually using the view orientation tools or the Middle Mouse Button to rotate the model.

6. - The first thing we need to do is draw a rectangle and center it about the origin. Select the "**Rectangle**" tool from the "Sketch" tab or by making a Right Mouse Button click in the graphics area, from the pop up menu. Make sure we have the "Corner Rectangle" option selected in the Rectangle's Property Manager.

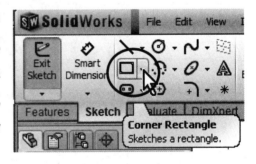

Click and drag in the graphics area to draw a Rectangle around the origin as shown. Left-Mouse-Click and drag to the opposite corner. Don't worry too much about the size; we'll dimension it in a later step.

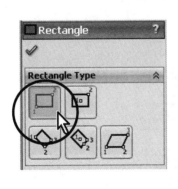
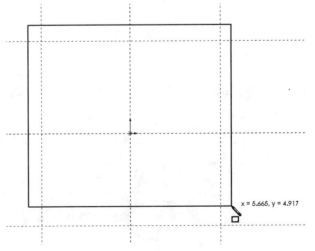

7. -   Notice the lines are in green after finishing the rectangle. The color green means the lines are selected. You can unselect them by hitting the Escape (Esc) key, this will also de-select (turn off) the rectangle tool. We only need one rectangle in this sketch, so hit the "Esc" key. Now we will draw a "**Centerline**" from one corner of the rectangle to the opposite corner. The purpose of this line is to help us center the rectangle about the origin. We'll also learn a faster way to do this in the next few steps. From the "Sketch" tab select the Line command's drop down arrow, and select "Centerline" (Menu "Tools, Sketch Entities, Centerline").

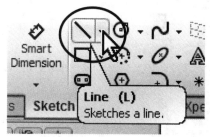
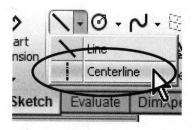

8. -   SolidWorks indicates that we will start or finish a line at an existing entity with yellow icons; when we locate the cursor near an endpoint, line, edge, origin, etc. it will "snap" to it. Click in one corner of the rectangle, click in the opposite corner as shown and press the "Esc" key to complete the Centerline command.

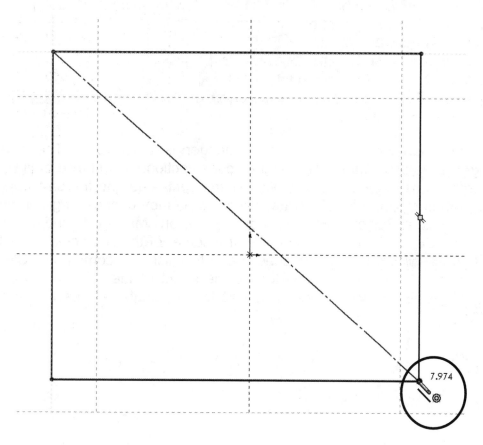

17

9. -   Next we want to make the midpoint of the centerline coincident to the origin; for this step we will add a **"Midpoint"** geometric relation between the centerline we just drew and the part's origin. Select from the menu "**Tools, Relations, Add**" or the "**Add Relation**" icon from the "Display/Delete Relations" drop-down icon. By adding this relation, the centerline's mid point will be forced to coincide with the origin; this way the rectangle (and the part) will be centered about the origin. Centering the part about the origin and the model planes will be useful in future operations.

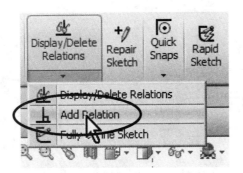

 "Add Relation" can also be accessed through the right mouse button menu, or be configured to be available in the **Mouse Gestures** shortcuts.

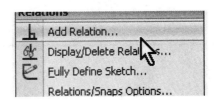

10. - The "Add Relations" Property Manager is displayed. The **Property Manager** is the area where we will make our selections and choice of options for most commands. Select the previously made centerline and the part's origin by clicking on them in the graphics area (notice how they change to green and get listed under the "Selected Entities" box). Click on "**Midpoint**" under the "Add Relations" box to add the relation. Now the center of the line coincides with the Origin. Click on OK (the green checkmark) to finish the command. Click and drag one corner of the rectangle to see the effect of the relation. Notice the center of the line stays in the origin and the rectangle resizes symmetrically about the center.

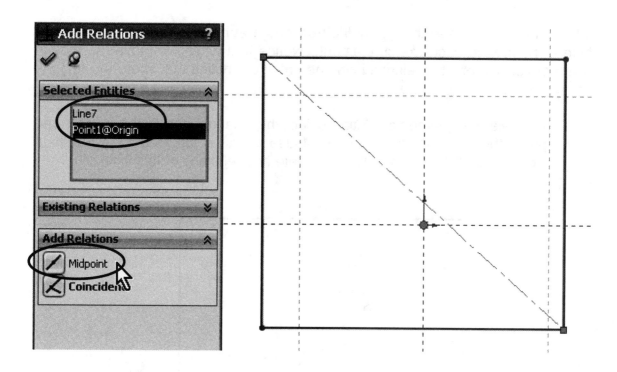

11. - We added a geometric relation manually, and we also added geometric relations automatically when we drew the rectangle and the centerline in the previous step. SolidWorks allows us to view the existing relations graphically between sketch elements. Go to the menu, "**View, Sketch Relations**" if not already activated, or from the "Hide/Show Items" drop down icon in the graphics area.

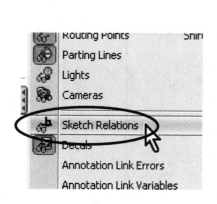

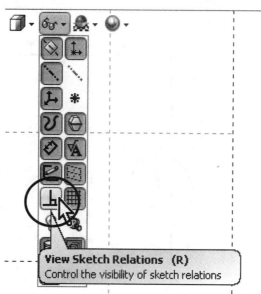

12. - Now we can see the geometric relations graphically represented by small blue icons next to each sketch element. Notice that when we move the mouse pointer over a geometric relation icon, the entity or entities that share the relation are highlighted.

 To delete a geometric relation select the relation icon in the screen and press the "Delete" key, or Right Mouse Click on the Geometric Relation icon and select "Delete". (Do not delete any relations at this time!)

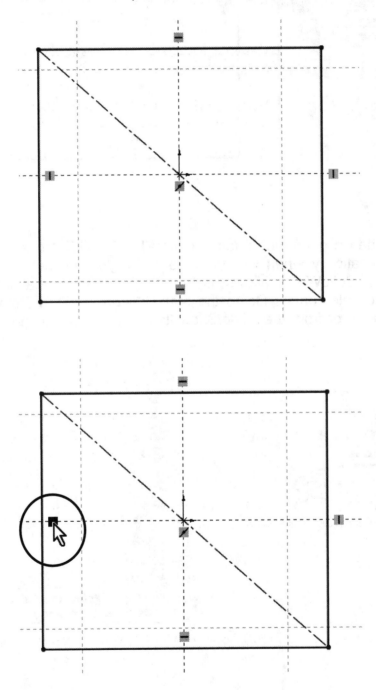

The **Sketch Origin** defines the local Horizontal (short red arrow) and local Vertical (long red arrow) directions for the sketch, this is important because we may be looking at the part in a different orientation, and vertical may not necessarily mean "up" on the screen; since we are working in three dimensions, and we can view the part from any view orientation; it is a convenient way for us to know where "up" is in the sketch. In SolidWorks we have the following basic types of geometric relations between sketch entities:

	Vertical Parallel to the sketch vertical direction (long red arrow in the origin)
	Horizontal Parallel to the sketch horizontal direction (short red arrow in the sketch origin)
	Coincident is when an endpoint touches another line, endpoint, model edge or circle.
	Midpoint is when a line endpoint coincides with the middle of another line or model edge. A Midpoint relation implies it is also Coincident.
	Parallel is when two or more lines or a line and a model edge have the same inclination.
	Perpendicular is when two lines are 90 degrees from each other, like vertical and horizontal lines. Note that the lines don't have to touch each other in order to be perpendicular.
	Concentric is when two arcs or circles share the same center. Concentric can be also between a point or line's endpoint and an arc or circle's center.
	Tangent is when a line and an arc or circle, or two arcs or circles are tangent to each other.
	Equal is when two or more lines are the same length, or two or more arcs or circles have the same diameter.
	Collinear is when two or more lines lie on the same line.

13. -   Once we have added the geometric relation "Midpoint", the next step is to dimension the rectangle. Turn off the geometric relations display in the menu **"View, Sketch Relations"** to avoid visual clutter in the screen if so desired. Click with the Right Mouse Button in the graphics area and select "**Smart Dimension**" or select the "Smart Dimension" icon from the Sketch Toolbar.
Notice the cursor changes adding a small dimension icon next to it. This icon will indicate the user the Smart Dimension tool is selected.

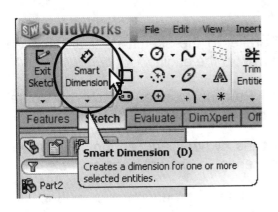
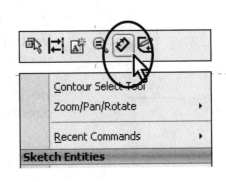

 Smart Dimension can also be activated from the Mouse Gestures (Right-Mouse-Click and drag up by default).

14. -   Adding dimensions in SolidWorks is simple and straight forward. Click to select the right vertical line and then click just to the right to locate the dimension. SolidWorks will show the "Modify" dialog box, where we can add the 2.625" dimension. Repeat with the top horizontal line and add a 6" dimension. As soon as the dimension value is accepted, the geometry updates to reflect the correct size.

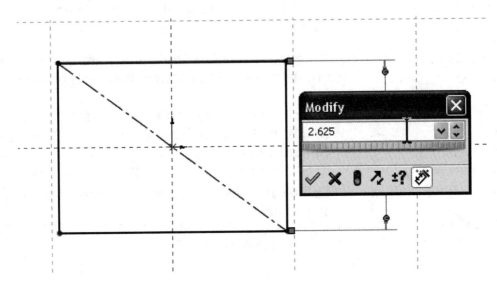

View the Appendix if you need to change the document's **units** from millimeters to inches or vice versa. You can also override the default units for one dimension by adding "in" or "mm" at the end of the value in the Modify dialog box.

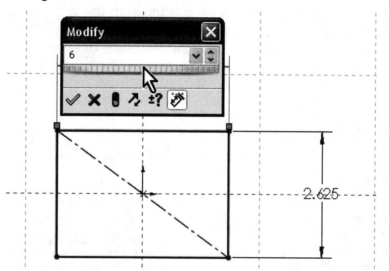

 To change a dimension's value double click on it to show the "Modify" box.

After dimensioning the lines, notice the lines changed from Blue to Black. This is the way SolidWorks lets us know that the geometry is defined, meaning that we have added enough information (dimensions and/or geometric relations) to define the geometry in the sketch. The status bar also shows "Fully Defined" in the lower right corner. This is the preferred state before creating a feature, since there is no information missing and the geometry can be accurately described.

A sketch can be in one of several states; the three main ones are:

- **Under Defined**: (BLUE) Not enough dimensions and/or geometric relations have been provided to define the sketch. Sketch geometry is blue and lines/ endpoints can be dragged with the left mouse button.

- **Fully Defined**: (BLACK) The Sketch has all the necessary dimensions and/or geometric relations to completely define it. <u>This is the desired state</u>. Fully defined geometry is black.

- **Over Defined**: (RED) Redundant and/or conflicting dimensions and/or geometric relations have been added to the sketch. If an over-defining dimension or relation is added, SolidWorks will immediately warn the user. If an over-defining geometric relation is added, delete it or use the menu, **"Edit,**

Undo" or select the "Undo" icon. If an over-defined dimension is added, the user will be offered an option to cancel it.

15. - Now that the sketch is fully defined, we will create the first feature of the housing; this is when we go from the 2D Sketch to a 3D feature. Click in the Features tab in the Command Manager and select the "**Extrude**" icon, or click in the "Exit Sketch" icon in the Sketch toolbar. In the second case SolidWorks remembers that we wanted to make an Extrusion in the first place, and displays the Extrude command's property manager. Notice that the first time we create a feature in a new part, SolidWorks changes the view orientation to an Isometric view and gives us a preview of what the feature will look like when finished.

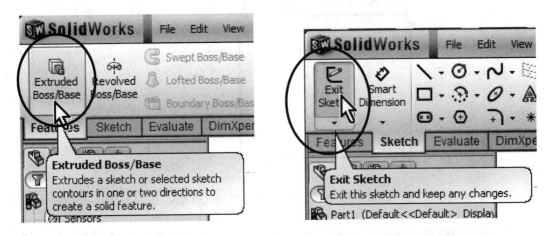

Select the options indicated in the "**Extrude**" command to make the extrusion 0.25" thick. To finish the command, select the OK button or press the "Enter" key.

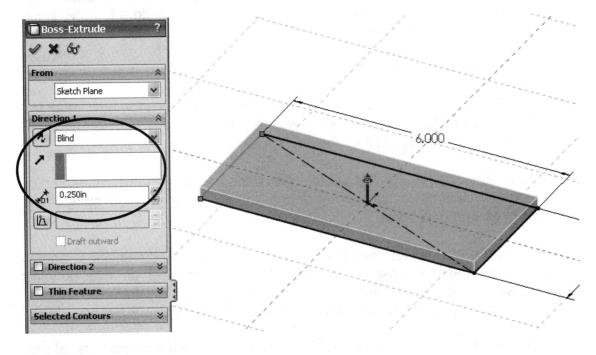

16. - After the first extrusion is completed, notice that "Extrude1" has been added to the Feature Manager. The confirmation corner is no longer active. The status bar now reads "Editing Part" to alert us that we are now editing the part and not a sketch.

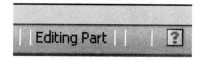

 Expanding the "Boss-Extrude1" feature in the Feature Manager by clicking on the "+" on the left side of it, we see that "Sketch1" has been absorbed by the "Boss-Extrude1" feature.

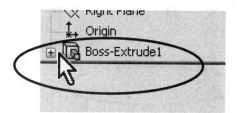
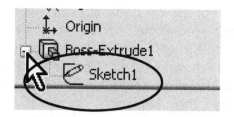

17. - The second feature will be similar to the first one but with different dimensions. To create the second extrusion, we need to make a new sketch. When we select the Extrude or Sketch icon, SolidWorks gives us a yellow message in the Property Manager asking us to select a Plane or a planar (flat) face. We'll select the top face of the previous extrusion for the next feature (If a Plane or flat face is pre-selected, the Sketch opens immediately in that Plane/face).

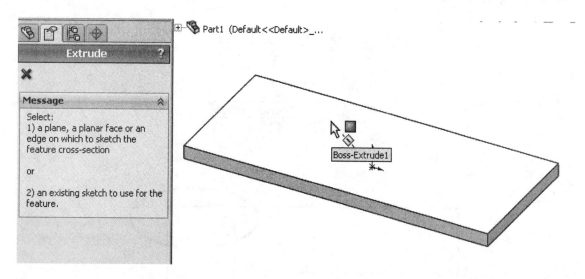

18. - Only in the first sketch of the part, the view is oriented to the sketch plane. To help us get oriented, we will switch to a **Top View** to see the part from the top. In SolidWorks the user is free to work in any orientation, as long as he/she is able to see what they are doing. Re-orienting the view helps new users get used to 3D in a more familiar way by looking at it in 2D. Use the "View Orientation" icon, or the default shortcut "Ctrl+5", as indicated in the tooltip.

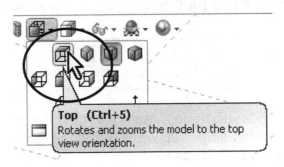

Top (Ctrl+5)
Rotates and zooms the model to the top view orientation.

19. - For the second extrusion, we'll use the "Center Rectangle" command. Click in the Rectangle's tool drop down menu in the Sketch tab, and select "**Center Rectangle**"; if you selected the rectangle as before, you can change the rectangle type to "Center Rectangle" from the Rectangle's Property Manager. To make the rectangle, click in the Origin first, then on the top edge of the first extrusion (or the bottom, it really doesn't

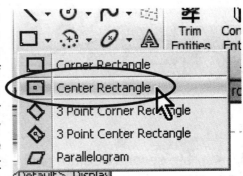

matter) to finish it. Notice the yellow Coincident icon as the pointer is in the origin and then on the model edge. By doing it this way, we automatically add coincident relations to the origin and the top edge. The "Center Rectangle" command saves us from adding the centerlines and midpoint relations making the rectangle centered about the origin in a single operation.

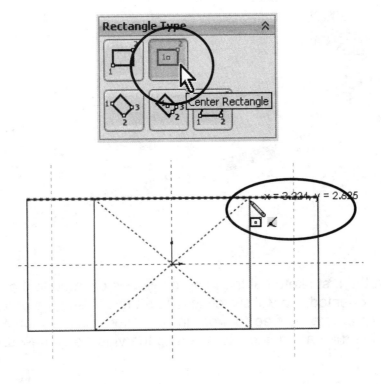

20. - Select the "Smart Dimension" tool (toolbar, right mouse menu or mouse gesture), and dimension the rectangle 4" wide by selecting the top line as shown. Adding this dimension will fully define the sketch.

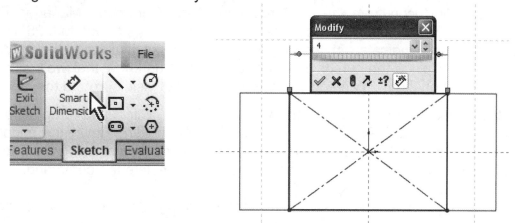

21. - We are now ready to make the second extruded feature. Select the "**Extrude**" command in the "Features" tab of the Command Manager (or "Exit Sketch" if you initially selected "Extrude") and make the extrusion 3.5". From the Standard Views icon, select the **Isometric** view (or the "Ctrl+7" shortcut or Mouse Gesture) to see the preview of the second extrusion. Click on OK to complete the command.

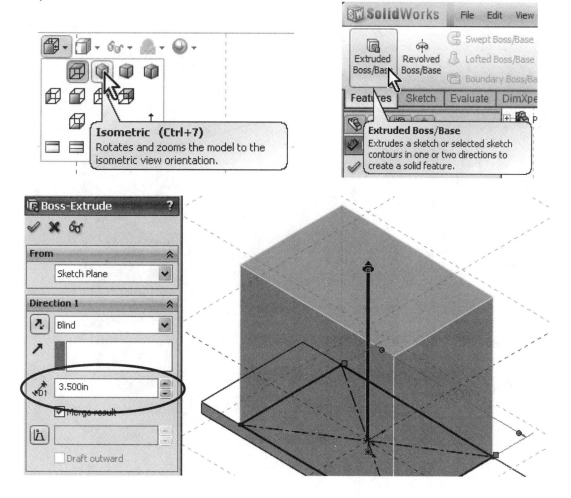

22. -   The next step is to round the edges of the two extrusions. To do this, we will select the **Fillet** command. The Fillet is what's called an applied feature; we don't need a sketch to create it, and it's applied directly to the solid model. Select the "Fillet" icon from the Features Tab of the Command Manager. By default, "Constant Radius" type is selected. We'll change the radius to 0.25" and select the corners indicated in the preview. SolidWorks highlights the model edges when we place the cursor on top of them to let us know that we'll select them if we click on them. If an edge to be selected is not visible, rotate the model using the menu "**View, Modify, Rotate**".   ⟳ Rotate   Click and drag in the graphics area to rotate. Another way to do this is by holding down the middle mouse button (scroll wheel), and dragging in the graphics area or using the arrow keys. Click OK when all eight edges are selected to complete the command.

   If an edge or face is mistakenly selected, simply click on it again to de-select it.

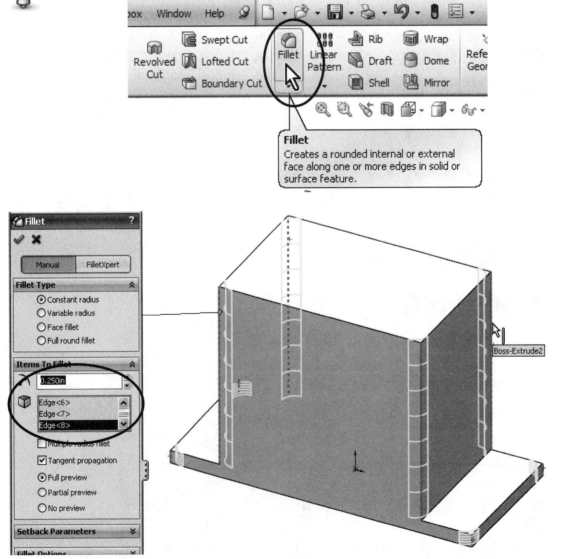

A handy feature is a "**Magnifying Glass**" to selectively zoom in on one area of the model. To activate it, use the default shortcut "G" in the keyboard. To make multiple selections with it, hold down the "Ctrl" key, otherwise, the Magnifying Glass will turn off after making one selection or after pressing "G" again. Scrolling with the mouse wheel will zoom inside the Magnifying Glass for more or less magnification.

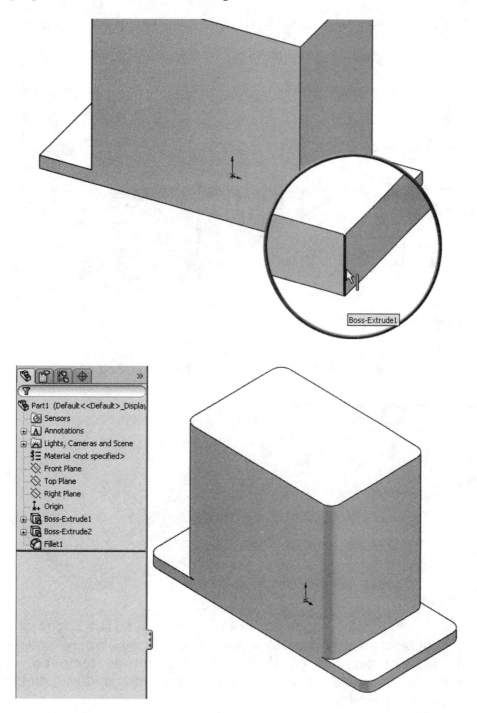

23. - Repeat the fillet command to add a 0.125" radius fillet at the base of the Housing.

By selecting a face SolidWorks rounds all the edges of that face.

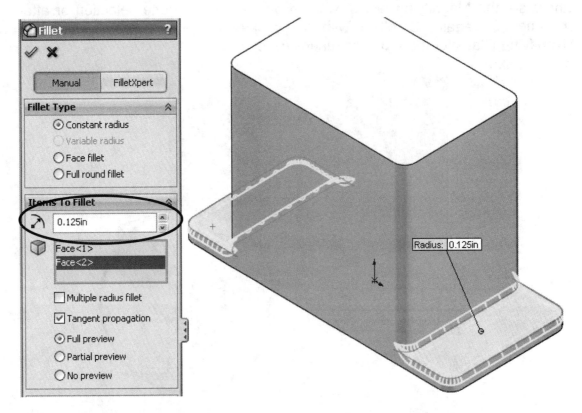

Notice how the new fillets blend nicely with the existing fillets.

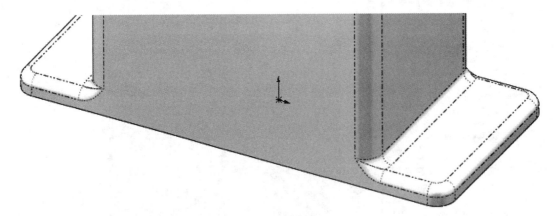

We can change the appearance of tangent edges (The edges where two tangent faces meet) by selecting the menu, "**View, Display**" and selecting the display option desired: **Visible**, as **Phantom** or **Removed**. Explore the different options to find the one you feel more comfortable with. In this book Phantom lines are used for clarity.

24. - We will now remove material from the model using the "**Extruded Cut**" command. Switch to a **Top View** using the View Orientation Toolbar, Mouse Gesture or shortcut Ctrl+5.

Select "**Extruded Cut**" from the Features tab; you will be asked to select a face or planar face just as with the "**Extruded Boss**". Select the top most face to create a new Sketch, and using the "**Corner Rectangle**" tool from the now visible "Sketch" toolbar, draw a rectangle *inside* the top face.

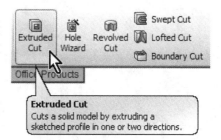

Extruded Cut
Cuts a solid model by extruding a sketched profile in one or two directions.

To add the dimensions select the "**Smart Dimension**" tool; we can add dimensions from sketch geometry to model edges simply by selecting them. Select a Sketch line, click on a model edge parallel to it, and finally click to locate the dimension in the screen. You will be asked to enter the dimension as before. Repeat to add the rest of the dimensions.

Smart Dimension

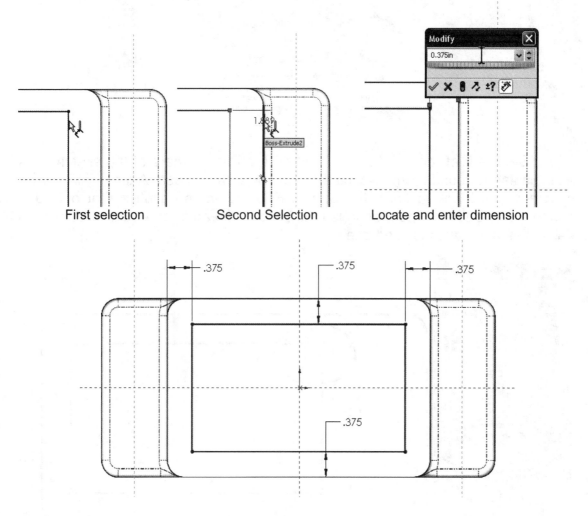

First selection   Second Selection   Locate and enter dimension

25. - If needed, switch to a "Hidden Lines Removed" mode from the **View Style** icon to view the model without shading to facilitate visualization.

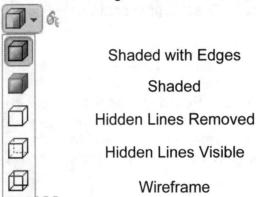

Shaded with Edges

Shaded

Hidden Lines Removed

Hidden Lines Visible

Wireframe

26. - In this feature, we will round the corners in the sketch using a **Sketch Fillet**. We can add the fillets as applied features like before, but in this step we chose to show you how to round the corners in the Sketch *before* making the "Extruded Cut" feature. Select the "**Sketch Fillet**" icon from the Sketch Toolbar.

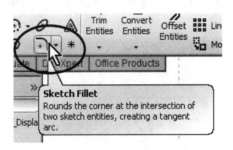

Set the fillet radius to 0.150", and click on the corners of the sketch lines as indicated to round them. Notice the preview in the screen. After clicking on all 4 corners, click OK to finish the Sketch Fillet command. Notice that only one dimension is added. The reason is that SolidWorks adds an equal relation from each fillet to the dimensioned one.

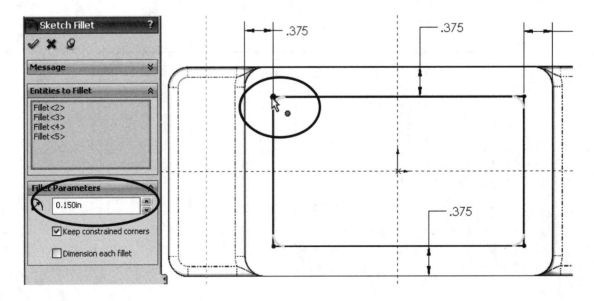

27. - Now we select the **Extruded Cut** icon from the "Features" tab in the Command Manager to remove the material. Opposite to the Boss Extrude feature that adds material, the Cut feature, as its name implies, removes material from the model.

 Change the view to Isometric....

...and **Shaded with Edges** for better visualization.

Make the cut 3.5" deep and click on OK to finish the cut.

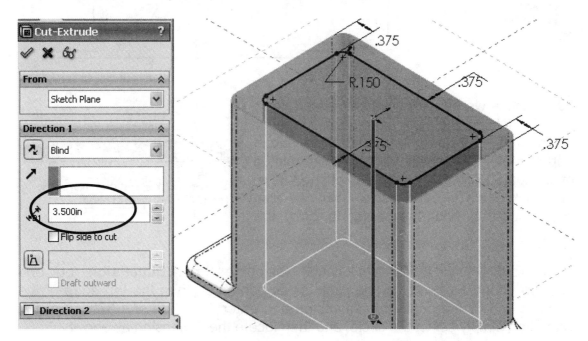

 Features can be **Renamed** in the Feature Manager for easier identification. To rename a feature slowly double-clicking the feature, or select and press F2 and type a new name (just like renaming files in Windows Explorer!).

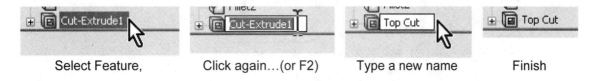

| Select Feature, | Click again...(or F2) | Type a new name | Finish |

28. - In the next step we will add a simple round boss to the front of the Housing. Switch to a **Front View** using the "View Orientation" toolbar, Mouse Gestures or shortcut Ctrl+1.

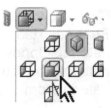

Select the "Sketch" icon from the "Sketch" tab in the Command Manager and click in the front face, or, the reverse order: select the face first, and then click in the "Sketch" icon. Notice that SolidWorks highlights the entire flat face before selecting it to add a sketch.

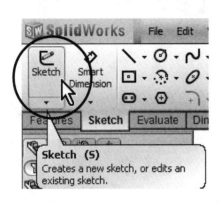

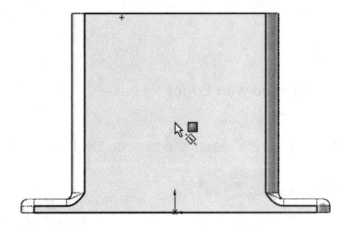

29. - Once we have the sketch created, from the "Sketch" tab of the Command Manager select the **"Circle"** tool. Draw a circle approximately as shown; click near the middle of the part to locate the center of the circle, and click again to set its size (You can also click and drag from the center to draw the circle).

To dimension the circle's location, select the **"Smart Dimension"** tool, and click on either the center of the circle or its perimeter, then the top edge of the housing and finally locate the dimension and enter the value of 1.875". For the Diameter, simply select the circle, and then locate the dimension as shown.

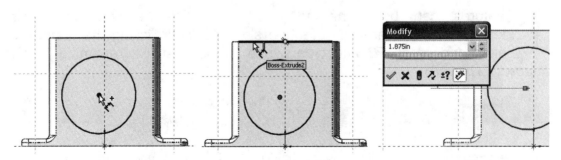

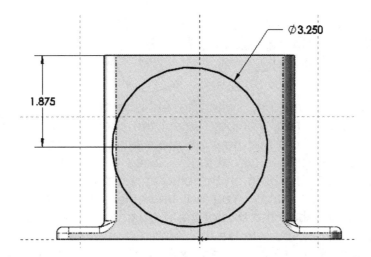

⌀3.250

1.875

30. - To locate the circle horizontally centered in the part, we will add a **Vertical Relation** between the center of the circle and the part's origin. SolidWorks allows us to align sketch elements to each other or to existing model geometry (edges, faces, vertices, planes, origin, etc.). From the Right Mouse button menu, select "**Add Relation**" (or the menu "**Tools, Relations, Add**"), Select the circle's center (Not the perimeter!) and the origin. Click on "**Vertical**" to add the relation and OK to finish. By adding this relation our sketch is now fully defined.

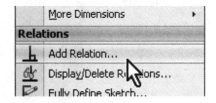

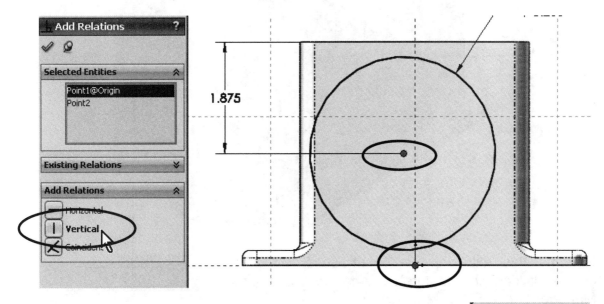

31. - After adding the **Vertical** geometric relation exit the sketch. We'll use a time saving feature called "**Instant 3D**" to make the extrusion. It should be active by default in the "**Features**" toolbar in the Command Manager, otherwise simply click to activate it.

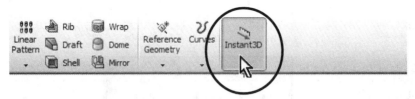

To make the extrusion, switch to an isometric view using the "View Orientation" Toolbar, select the circle of the previous sketch. Take note that now we are editing the part, we left the sketch editing environment. Once the sketch circle is selected, click and drag on the blue arrow, this is the handle to extrude the sketch. You will immediately see a dynamic ruler that will show the size of the extrusion as you drag it. Make sure to extrude it 0.250". When you release the handle, a new extrusion will be created. To modify this extrusion, simply select a face of this extrusion and drag on the handle to the new size.

 You can control the size with more precision by dragging the handle over the ruler's marks; this way the handle will snap to them.

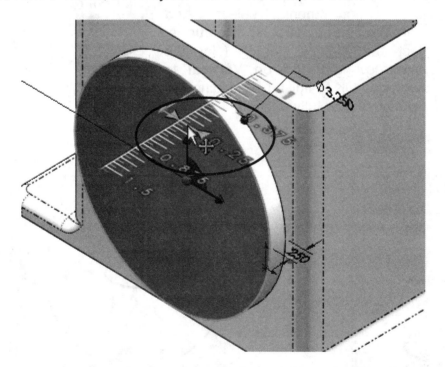

Rename this extrusion as "Front Boss" by slowly double-clicking the feature's name in the Feature Manager.

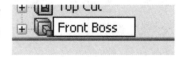

32. - The next step is to create an identical extrusion on the other side of the Housing. To make it we'll use the "**Mirror**" command. This will make an identical 3D copy of the extrusion we just made. Switch to an Isometric view to help us visualize the Mirror's preview and make sure we are getting what we want.

33. - Select the "**Mirror**" icon from the Features tab in the Command Manager.

34. - From the Mirror's Property Manager, we have to make two selections. The first one is the Mirror Face or Plane and the second is the feature(s) we want to make a mirror from. The face or plane that will be used to mirror the feature has to be in the middle between the original feature and the desired mirrored copy. Making the first extrusion centered about the origin causes the Front plane to be in the middle of the part, making it the best option for a Mirror Plane.

 To select the Front plane (make sure the "Mirror Face/Plane" selection box is highlighted, this means it is the active selection box), click on the "+" sign next to part's name to reveal a **fly-out Feature Manager**, from where we can select the Front Plane.

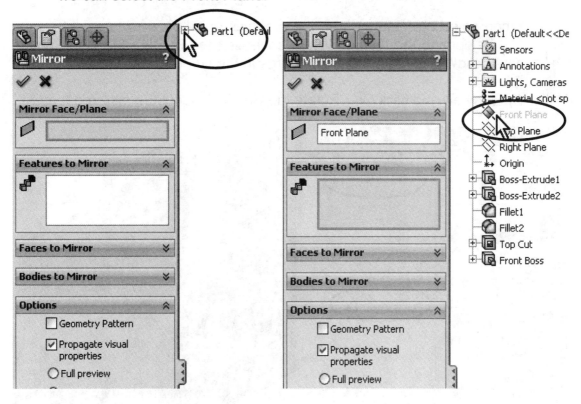

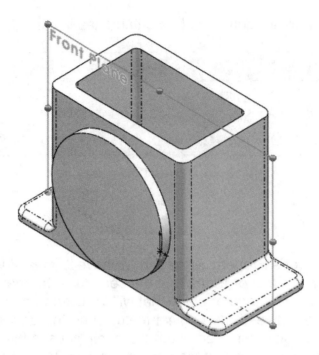

35. - After selecting the Front Plane from the fly-out Feature Manager, SolidWorks automatically activates the "Features to Mirror" selection box (now highlighted) and is ready for us to select the feature(s) we want to mirror. If the "Front Boss" extrusion was pre-selected, it will be listed automatically; otherwise we will select it either from the Feature Manager or in the graphics area.

 When selecting features from the graphics area, be sure to select a face that belongs to the feature. When traversing the Feature Manager, notice how SolidWorks highlights the features in the screen before selecting it.

Notice the preview and click OK. Rotate the view to inspect the mirrored feature.

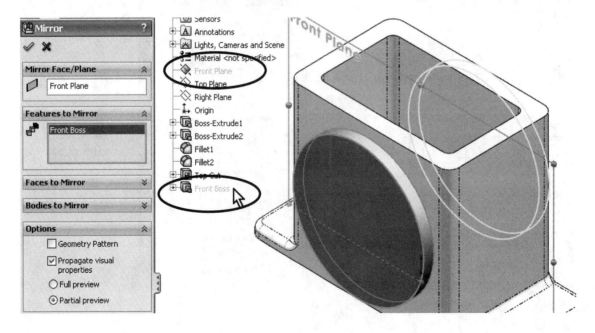

36. - In the next step we'll add the small boss at the right side of the Housing. Switch to a **Right View,** using the View Orientation toolbar (Shortcut Ctrl+4), and select the "**Sketch**" icon from the Command Manager's Sketch tab.

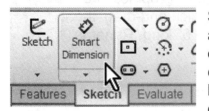

Select the rightmost face to create the Sketch, draw a circle using the "Circle" tool and add the dimensions shown. Just as we did with the front cylindrical boss, we'll add a **Vertical Relation** between the center of the circle and the part's origin.

From the Right Mouse click menu, select "**Add Relation**", select the center of the circle and the origin, and click on "Vertical" in the "Add Relations" box.

Select Sketch plane   Draw & dimension circle   Add vertical relation

When adding relations, if you accidentally select more entities than needed, you can unselect them in the screen, or select them in the selection box and remove them using the "Delete" key.

37. - Now we are ready to extrude the sketch to make the side boss. We'll use the "**Instant 3D**" function as we did in the previous extrusion. Exit the Sketch by selecting the "**Exit Sketch**" icon in the Command Manager's Sketch tab and change to an Isometric view. In the graphics area select the circle of the sketch we just drew, and click-and-drag the arrow along the ruler markers to make the extrusion 0.5" long. Rename this extrusion "Side Boss" in the Feature Manager.

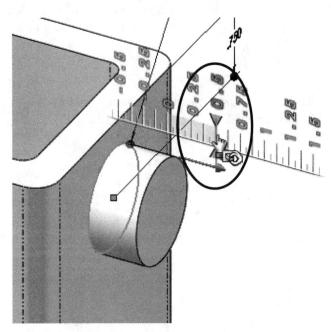

38. - Just like we did with the front circular boss, we'll mirror this extrusion about the **Right Plane** (which is also in the middle of the part). Select the "**Mirror**" command from the "Features" tab in the Command Manager, and using the fly-out Feature Manager select the "Right Plane" as the "Mirror Face/Plane" and the "Side Boss" extrusion in the "Features to Mirror" selection box to complete the Mirror command.

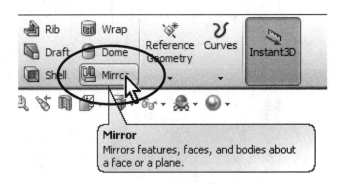

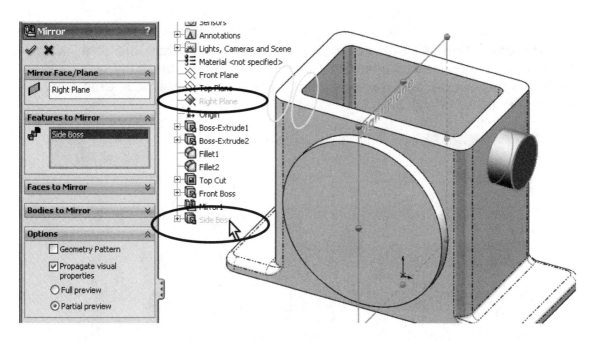

39. - We'll now make the circular cut in the front of the Housing. Change to a "Front View" for easier visualization. Select the "**Sketch**" icon from the Sketch tab and click in the round front face of the part.

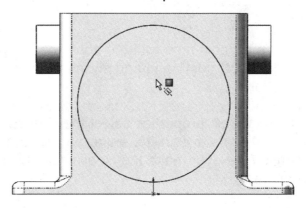

Draw a circle using the "**Circle**" tool and dimension it 2.250" diameter.

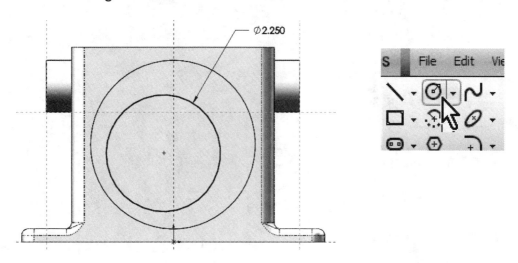

To locate the circle in the center of the circular face, we'll add a "**Concentric Relation**". Select the "**Add Relation**" icon from the Right Mouse button menu, and select the circle we just drew and the edge of the circular face. Click "**Concentric**" to add the relation and center the circle. Click OK to finish the command.

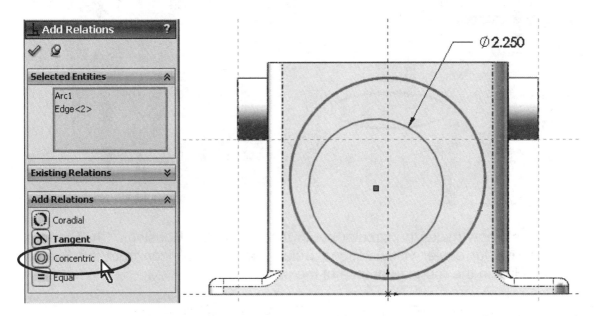

40. - Now that the circle is concentric with the boss, we'll make the cut. Select the "**Extruded Cut**" command and switch to an isometric view for better visualization.

From the "**Extruded Cut**" properties select the "**Through All**" option; this way the cut will go through the entire part regardless of its size. In other words, *if* we ever make the Housing wider, the cut will still go through the part.

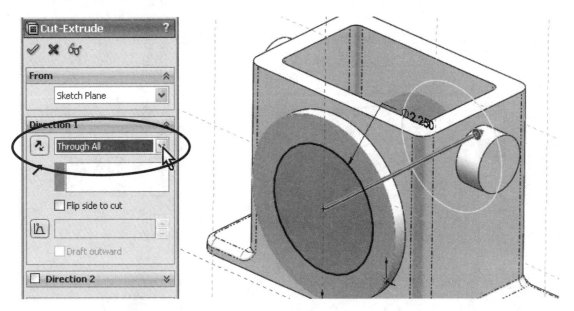

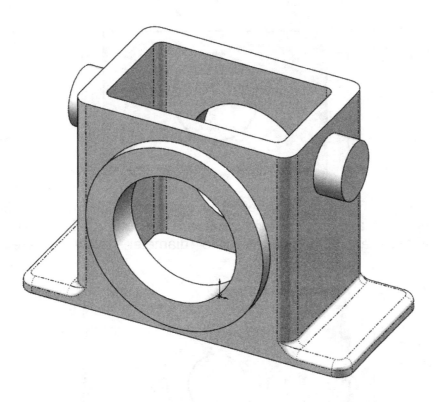

41. - We will now make a hole in the boss added in step 37 for a shaft. Switch to a "**Right View**" and create a sketch on the small circular boss' face by selecting the "**Sketch**" icon and then the circular face to locate it.

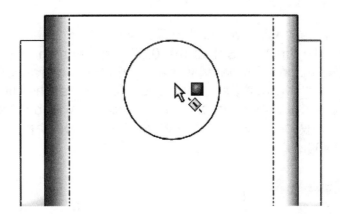

We know that we want the hole to be concentric with the boss. In order to do this we can draw the circle and add a concentric relation as we did before; however, this is a two step process. Instead, we will do it in one step as follows: Select the "**Circle**" tool icon from the "Sketch" toolbar and *before* drawing the circle, move the cursor and <u>rest it</u> on top of the circular edge as shown, the center of the circular edge is revealed in a fraction of a second. DO NOT CLICK ON THE EDGE. This highlight works only if you have a drawing tool active (Line, Circle, Arc, etc.). This technique can be used to reveal any circular edge's center.

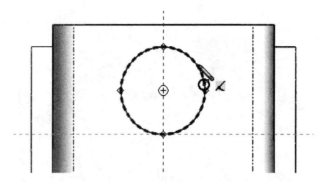

42. - Once the circular edge's center is revealed, click in the center to start drawing the circle and automatically capture a concentric relation with the boss. Finish the circle and dimension it 0.575" diameter. Now the sketch is fully defined.

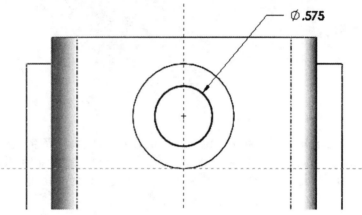

43. - Since this hole will be used for a shaft, we need to add a bilateral **tolerance** to the dimension. Select the 0.575" dimension in the graphics area, and from the dimension's Property Manager, under "Tolerance/Precision" select "Bilateral". Now we can add the tolerances. Notice that the dimension changes immediately in the graphics area. This tolerance will be transferred to the Housing's detail drawing later on. If needed, tolerances can also be added later in the detail drawing.

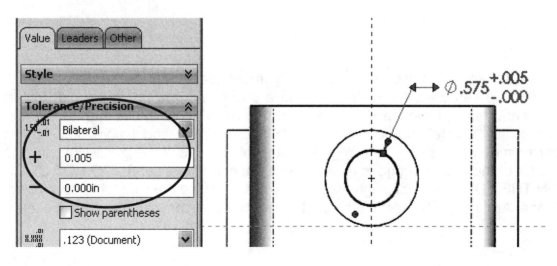

44. - Now we can make a Cut with the "Through All" option using the "**Extruded Cut**" command. Rename the Feature "Shaft hole".

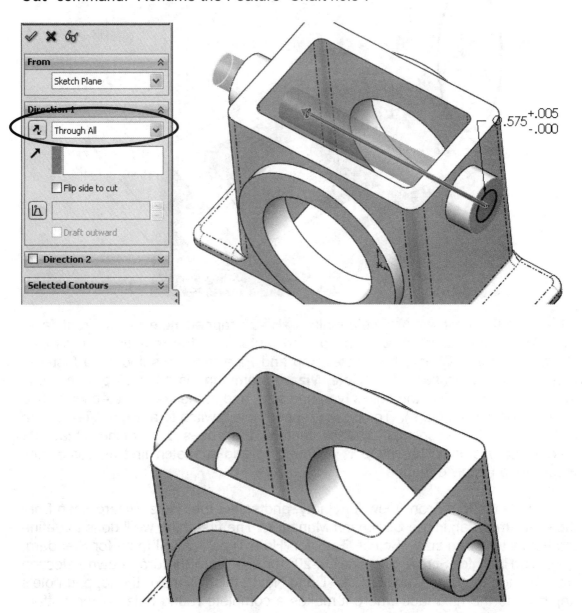

45. - Another way to do this cut is using the "Instant 3D" feature. To use "Instant 3D" to make a cut, click in the "**Exit Sketch**" icon, change to an Isometric view for clarity, and just like we did for the Extrusion, select the sketch circle and click-and-drag the handle *into* the part. You'll see how the part is cut as the arrow is dragged. The only disadvantage to making the cut using this technique is that the "Through All" option is not available while dragging.

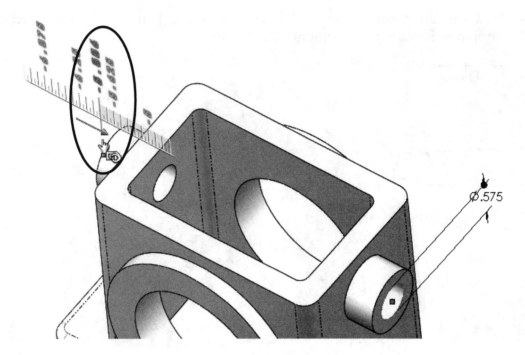

Ø.575

46. - For the next feature we'll make a ¼"-20 tapped hole in the front face. SolidWorks provides us with a tool to automate the creation of simple, Countersunk and Counterbore holes, tap and Pipe taps by selecting a fastener size, depth and location. The "**Hole Wizard**" command is a two step process: in the first step we define the hole's type and size, and in the second step we define the location of the hole(s). To add the tapped hole, switch to a "Front View". The Hole Wizard is a special type of feature that uses 2 sketches that are automatically created, so there is no need to add a sketch first; it works very much like an applied feature.

Change to a Front View for clarity, and select the "Hole Wizard" icon from the "**Features**" tab in the Command Manager. The first thing we'll do is to define the hole's type and size. Select "**Tap**" for "Hole Type", "ANSI Inch" for Standard, "Tapped Hole" for Screw type and "¼-20" for size from the drop down selection list. Change the "End Condition" to "Up to Next", this will make the tapped hole's depth up to the next face where it makes a complete round hole. At the bottom activate the button to add **Cosmetic Threads.**

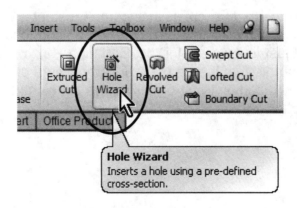

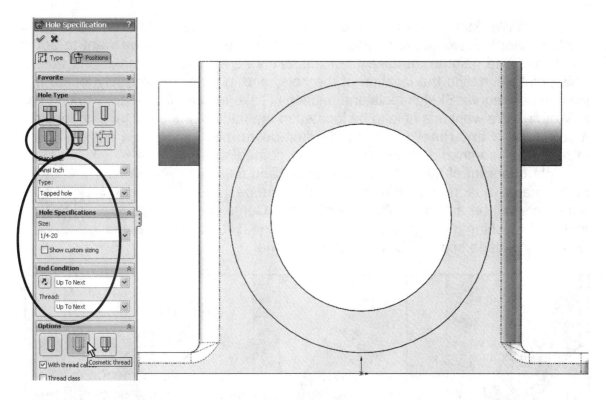

47. - The second step of the "Hole Wizard" is to define the hole's location. After selecting the "Positions" tab at the top of the properties, SolidWorks will ask to select a flat face to locate the hole(s). Select the round face at the front of the part.

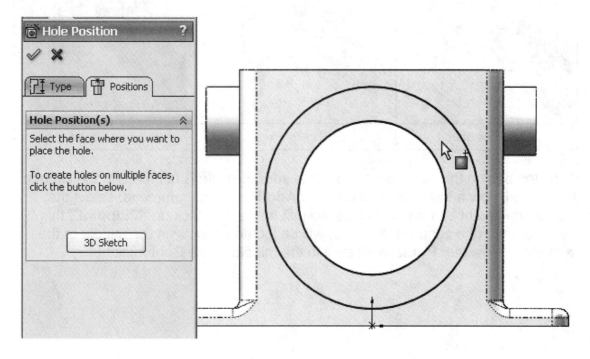

SolidWorks will add a "**Sketch Point**" where we select the face, and the Sketch Point tool will be automatically activated; this is in case we want to add more than one point in the face. For this exercise we will only make one hole. This point will define the location of the hole, and in order to precisely locate it, we can use regular Sketch tools and relations (Take notice that we are working in a Sketch). We want this hole to be located in the middle of the flat face, to locate it select the "**Centerline**" tool from the drop-down menu in the "Line" command. Start drawing the centerline at the right quadrant of the *outer circular edge*, and finish it in the same quadrant of the *inner circular edge*. The quadrants will be activated after selecting the Centerline tool, and touching (not clicking!) the circular edge. Hit the "Esc" key once to end the "Centerline" command.

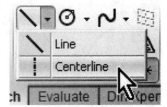

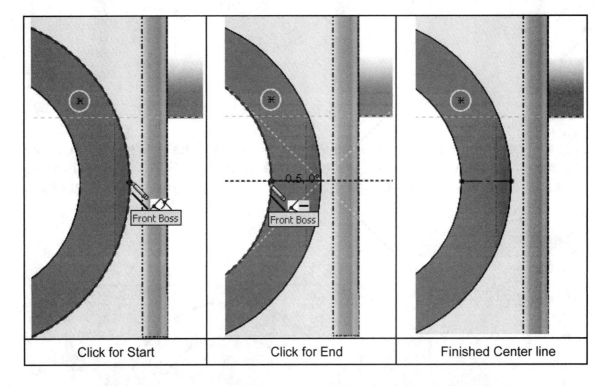

| Click for Start | Click for End | Finished Center line |

The idea behind this technique is to make sure the hole is centered in the circular face. We'll now add a relation using the "**Add Relation**" command; select the pre-existing point and the centerline we just made and click on "**Midpoint**", this way the point (the center of the hole) will be located exactly in the middle of the centerline. Now the "**Point**" is located in the middle of the Centerline.

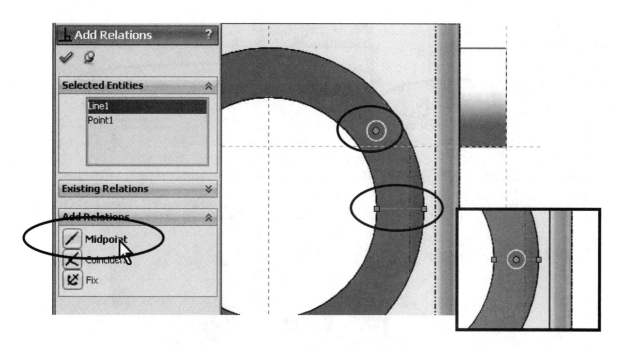

Click OK to close the "Add Relations" dialog, and then click OK again to finish the Hole Wizard.

 A quicker way to add this relation is to Window-Select the "**Point**" and the "**Centerline**", and from the pop up toolbar select "**Midpoint**" to add the relation. This approach works well when we pre-select the entities to relate.

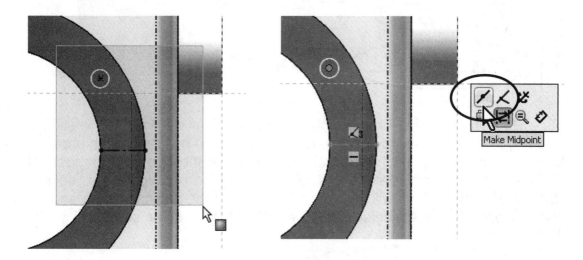

 In order to view the "Cosmetic Threads" (for a nicer look), Right-Mouse-Click in the "Annotations" folder at the top of the Feature Manager, select Details and activate the options "Cosmetic Threads" and "Shaded Cosmetic Threads". This is just for show, as there are no real threads modeled. It can be done, but it's mostly unnecessary in this type of design.

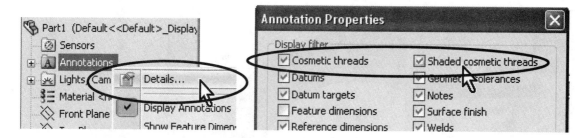

This is the finished ¼"-20 Tapped Hole with Cosmetic Threads.

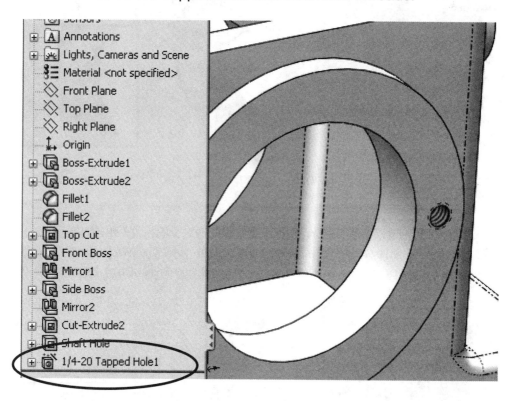

48. - After making the Tapped hole, we realize that we want the walls of the Housing to be thinner, and need to make a change to our design. In order to do this, we find the feature that we want to modify in the Feature Manager ("Top Cut") or in the graphics area, and select it. From the pop up toolbar, select the **"Edit Sketch"** icon. This will allow us to go back to the original sketch and make changes to it. Notice the selected feature is highlighted in the screen.

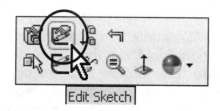

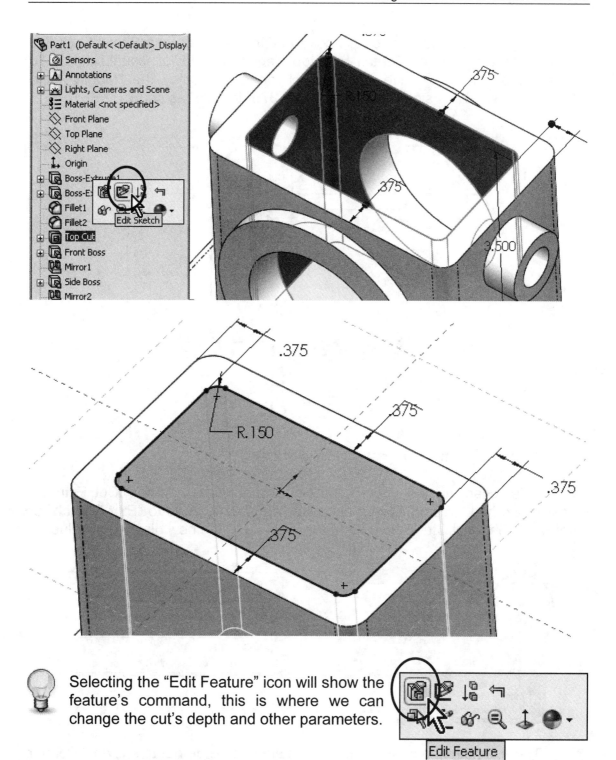

 Selecting the "Edit Feature" icon will show the feature's command, this is where we can change the cut's depth and other parameters.

Edit Feature

If we select the feature with a Right-Mouse-Click, we will see the pop up toolbar along with an options menu. The most commonly used commands are already in the pop up toolbar

 If "Instant 3D" is activated, selecting a feature will show its dimensions on the screen.

49. - What we just did was to go back to editing the feature's Sketch, just like when we first created it. Switch to a top view if needed for visualization. To change a dimension's value, double click on it to display the "Modify" box. Change the two dimensions indicated from 0.375" to 0.25" as shown.

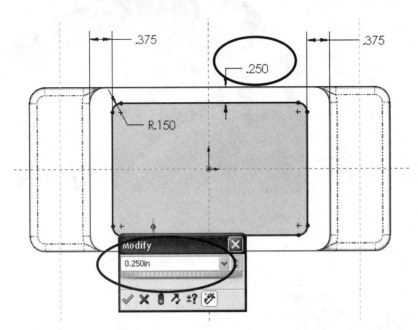

 To move a dimension just click-and-drag it.

50. - After changing the dimensions we cannot make another "Cut Extrude" because we had already made a cut; what we have to do is to **"Exit Sketch"** or **"Rebuild"** (Shortcut Ctrl+B) to update the model with the new dimension values.

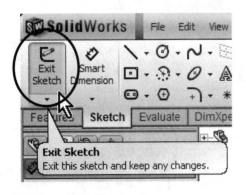 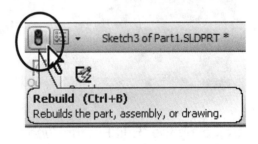

 There is no real purpose to this dimensional change but to show the reader how to change an existing feature's sketch if needed, including dimensions, geometric relations and add or remove sketch geometry.

Another way we can make this change is using the **"Instant3D"** functionality. The way it works is very simple: We select the feature that we want to modify either in the Feature Manager or the graphics area (in this case one of the inside faces which were made with the

Cut Extrude) and click-and-drag the **blue dots** at each of the dimensions that need to be modified until we get the desired value, without having to edit the sketch. Dragging the mouse pointer over the ruler markers will give you values in exact increments. Depending on the speed of your PC and the feature being modified Instant3D may be slow, as the solid model is dynamically updated.

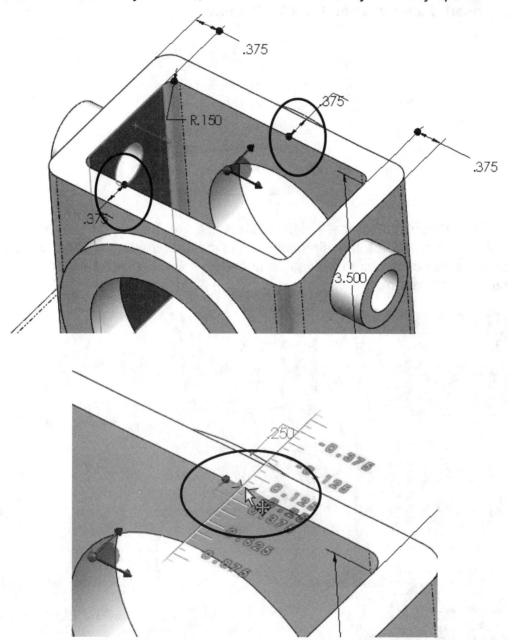

 A third option to change the dimensions is to select the value and type a new one. If Instant3D is not active, the Modify box will be displayed, and after making all the dimensional changes needed the user needs to select the Rebuild icon to update the model.

51. - Now we will add more tapped holes to complete the flange mounting holes. We'll use the first hole as a "seed" to make copies of it using the "**Circular Pattern**" command. In the "Features" tab, select the drop-down list below the "**Linear Pattern**" to reveal the drop down menu and select "**Circular Pattern**". Notice that commands are grouped by similar functionality. Optionally use the menu "**Insert, Pattern/Mirror, Circular Pattern**".

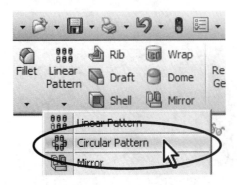

52. - The "**Circular Pattern**" needs a circular edge, a cylindrical surface or an axis as a reference for the direction of the circular pattern. Click inside the Parameters selection box to activate it, and then select the edge indicated for the direction of the pattern.

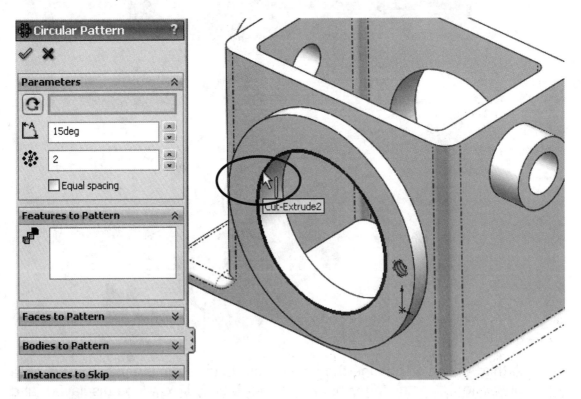

53. - Any circular edges or cylindrical faces that share the same axis can be used for a direction, as shown in the following images.

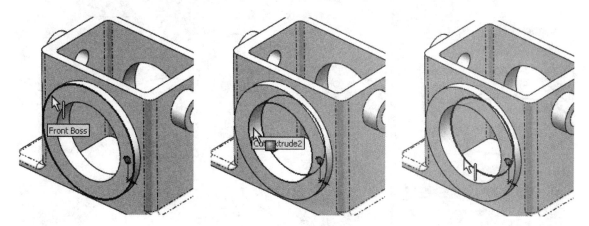

 Another option for direction is a temporary axis. Every cylindrical surface has a "Temporary Axis" that runs through its center. To view them select the menu "**View, Temporary Axes**" or in the "Hide/Show Items" toolbar.

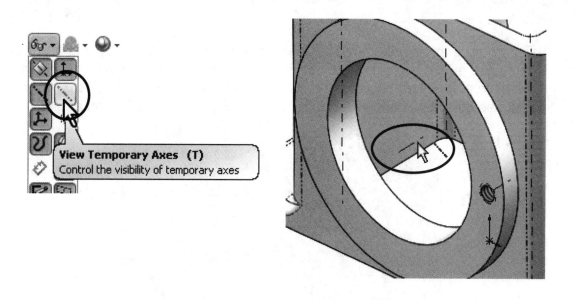

54. - Now click inside the "Features to Pattern" selection box to activate it (Notice it gets highlighted). Select the "¼-20 Tapped Hole1" feature from the fly-out Feature Manager; change the number of copies to six (notice this value includes the original), and make sure the "Equal spacing" option is selected to equally space the copies in 360 degrees. Notice the preview in the graphics area and click OK to finish the command.

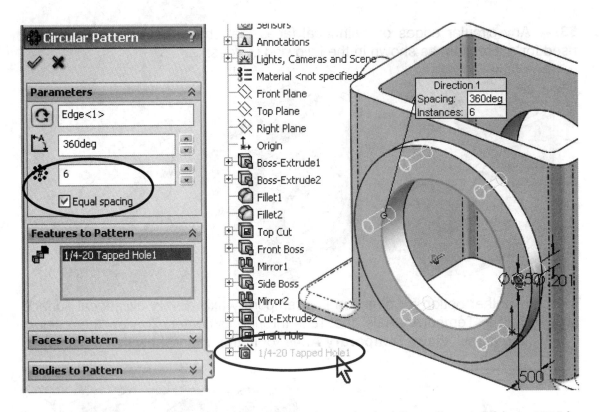

 The feature to be patterned can also be selected from the graphics area; in this case a face of the feature needs to be selected. Sometimes a face can be difficult to select because it may be small, like this hole. In this case, we can use the "**Magnifying Glass**" (Shortcut "G") to make selection easier.

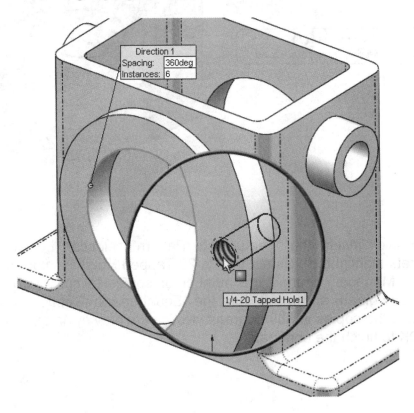

55. - Since we need to have the same six holes in the other side of the Housing, we will use the "Mirror" command to copy the Circular Pattern about the "Front Plane". Make this mirror about the Front Plane and mirror the "CircPattern1" feature created in the previous step.

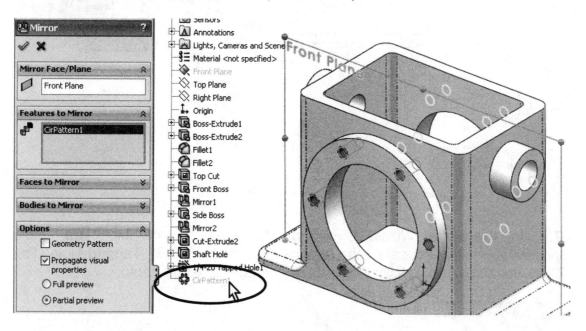

After the mirror, your part should look like this:

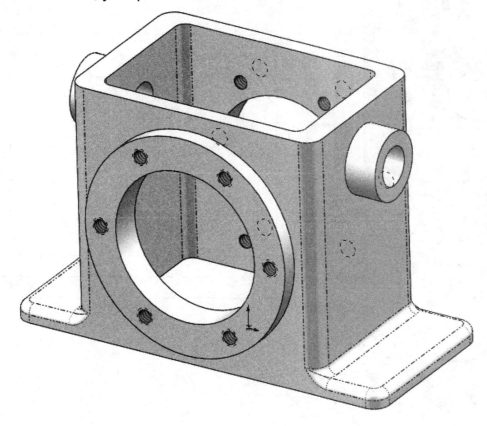

56. - We will now add four #6-32 tapped holes to the topmost face using the **Hole Wizard**. Switch to a Top View (Ctrl+5 or Mouse Gestures) for visibility, and select the "Hole Wizard" icon.

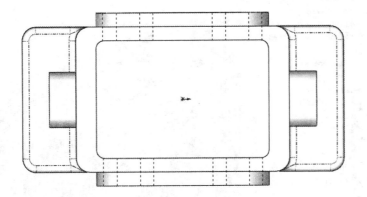

57. - In the Hole Wizard's Property Manager, select the "**Tap**" Hole Specification icon, and select the options shown for a #6-32 Tapped Hole. The "Blind" condition tells SolidWorks to make the hole an exact depth.

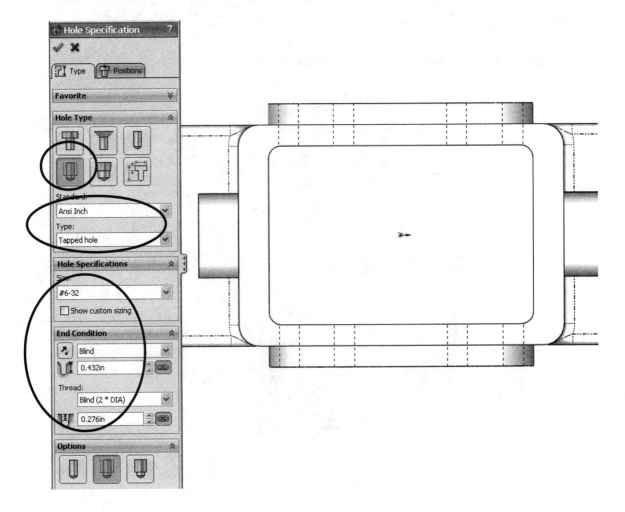

58. - Click in the "Positions" tab to define the hole's location. Select the top face where we need to add the holes, notice that when we select the face a "**Sketch Point**" is added, and now we are editing a sketch. Notice the Sketch Toolbar is automatically activated and the "Point" tool is pre-selected.

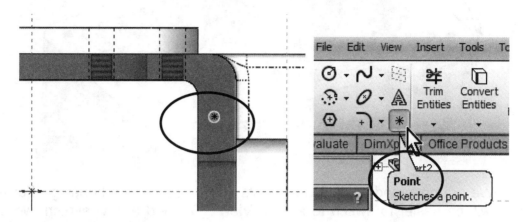

59. - With the "**Point**" tool active, touch each of the round corner edges to reveal their centers. Then click in their centers to add a point in each one; this way we'll make the points concentric to each corner fillet's center. If you have a point where we first selected the face, turn off the Point tool, select the point and delete it. Your model should look like the next image. Click OK to finish the Hole Wizard.

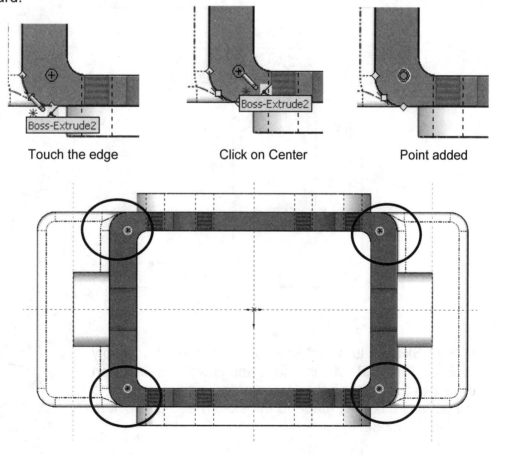

Touch the edge	Click on Center	Point added

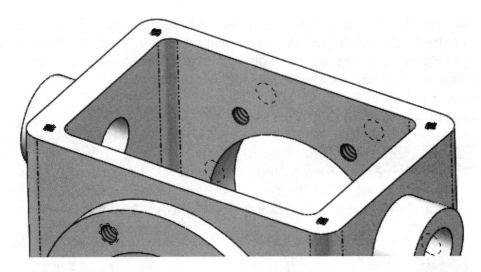

60. - We are now ready to make the slots at the base of the housing. For this task it will be easier to switch to a "Top View". To add a new sketch, we can select the "Sketch" command and click on the selected face as before, but in this case we'll learn how to use the pop up toolbar. Simply select the face, and from the pop up toolbar, select "Sketch". Notice there are two similar icons, the one on top is "**Edit Sketch**" to modify the sketch of the feature we selected, and the one below is "**Sketch**" to create a new sketch on the selected flat face.

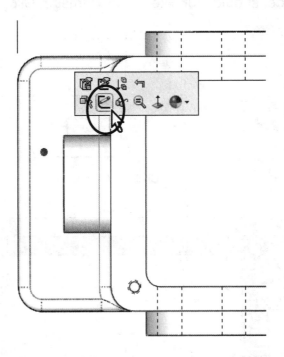

If "Edit Sketch" is selected instead of "Sketch", simply click on the red "X" in the transparent confirmation corner in the upper right corner of the graphics area to cancel any changes made to the sketch and go back to editing the model.

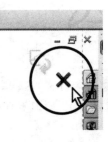

To make the slot, we'll use the **"Straight Slot"** command from the "Sketch" tab in the Command Manager. This tool will create a slot by first drawing the centerline and then defining the width of the slot. Select the **"Straight Slot"** icon and activate the "Add Dimensions" option, it will automatically add dimensions when we finish. The process is click to locate the center of one arc, click to locate the second, and click a third time to define the width. When finished, double click the dimensions to change them and make the slot 0.375" long and 0.250" wide.

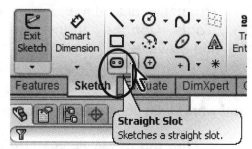

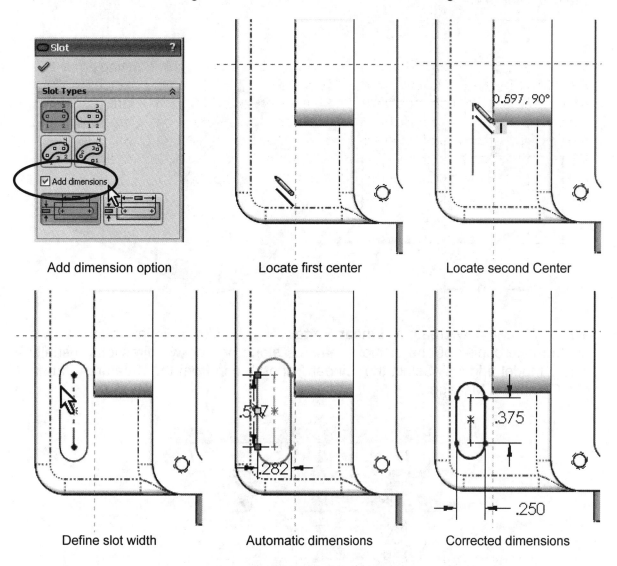

Add dimension option	Locate first center	Locate second Center

Define slot width	Automatic dimensions	Corrected dimensions

The "Slot" command has more options, including arc slots and overall slot length dimension.

To enable auto dimension while adding sketch elements select the menu "**Tools, Options**". In the "System Options" tab select the "Sketch" section and activate:

☑ Enable on screen numeric input on entity creation

This way when any sketch entity is created it will be automatically dimensioned and works with lines, rectangles, circles and arcs.

After the slot is defined, locate the bottom of the slot by adding two 0.5" dimensions from the lower and left edges of the base as shown. Finish the slot by making an "**Extruded Cut**" using the "Through All" option. Rename this feature "Slot".

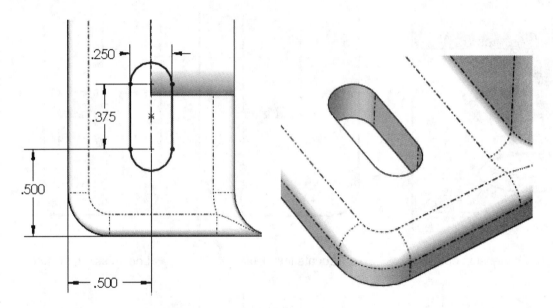

61. - We will now create a "**Linear Pattern**" of the slot. A linear pattern allows us to make copies of one or more features along one or two directions (usually along model edges). Select the "Linear Pattern" icon from the "Features" tab in the Command Manager.

Note: Using the Mirror command keeps the design intent better, but we chose to show the user how to use the Linear Pattern function instead.

62. - In the Linear Pattern's Property Manager, the "Direction 1" selection box is active; select the edge indicated for the direction of the copies. The copies will follow this direction. Any linear edge can be used as long as it is in the desired direction of the pattern.

Once the edge is selected, a grey arrow indicates the direction that the copies will be made. If the Direction Arrow in the graphics area is pointing in the wrong direction, click on the **"Reverse Direction"** button next to the "Direction 1" selection box.

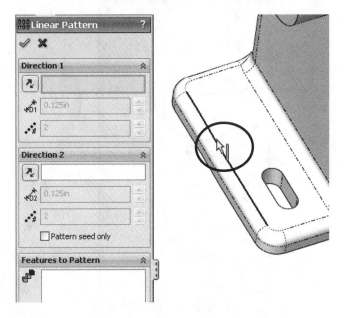

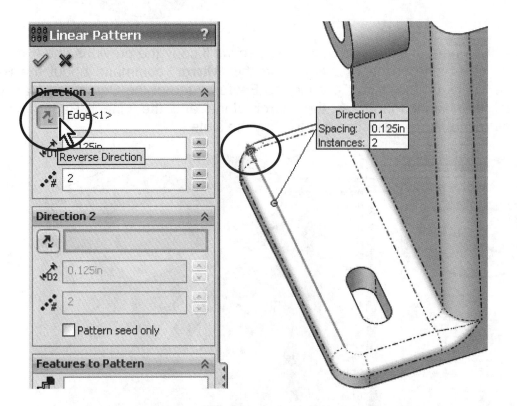

63. - Now click inside the "Features to Pattern" selection box to activate it and select the slot feature either from the fly-out Feature Manager or the graphics area. Change the spacing between the copies to 1.25" and total copies to 2. This value includes the original just like in the Circular Pattern. Click OK to finish the command.

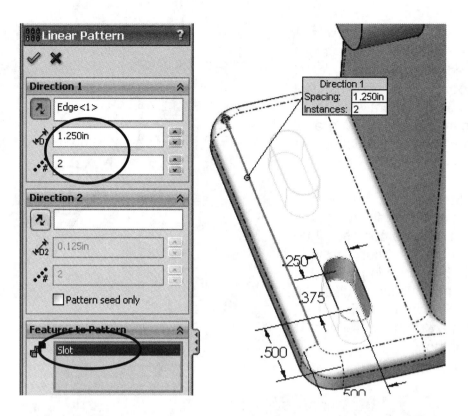

64. - We need the slots on both sides, so we'll copy the previous linear pattern to the other side of the Housing using the "**Mirror**" command about the Right Plane. Click on the "Mirror" icon in the Features tab of the Command Manager; select the "Right Plane" as the mirror plane and the Linear Pattern in the "Features to Mirror" selection box to copy the slots.

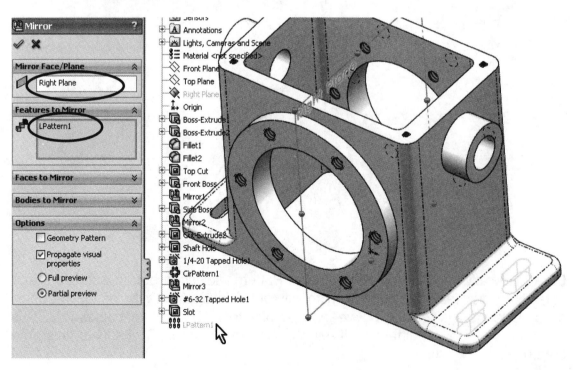

Selecting the Linear Pattern feature for the mirror also includes the "seed" feature (Slot).

65. - Using the "**Fillet**" command from the "Features" tab, add a 0.125" radius fillet to the edges indicated as a finishing touch. Rotate the model using the middle mouse button and/or change the display style to "Hidden Lines Visible" mode to make selection easier.

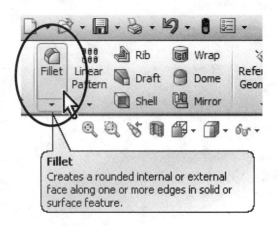

Fillet
Creates a rounded internal or external face along one or more edges in solid or surface feature.

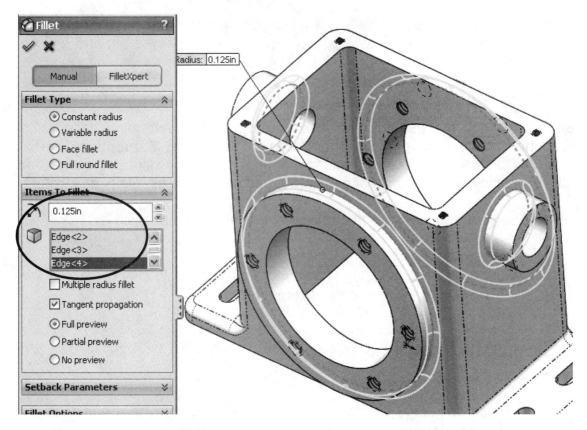

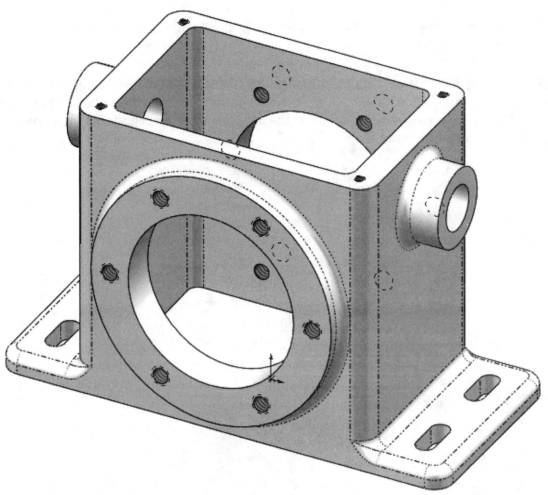

66. - Now that the model is finished, we can easily determine its physical properties, such as **Weight, Volume, Center of Mass and Moments of Inertia**. SolidWorks includes a built in materials library with many different metals and alloys, plastics, woods, composite materials and others like air, glass and water.

The library includes mechanical and thermal properties such as:

- Mass density
- Elastic and Shear modulus
- Tensile, Compressive and Yield strengths
- Poisson's ratio
- Thermal expansion coefficient
- Thermal conductivity
- Specific heat

 These properties are used by SolidWorks to determine a part's weight, or determine with SimulationXpress, the built in structural analysis software, if a component will fail under a given set of loading conditions.

To assign a material to a component, Right-Mouse-Click in the "Material" icon at the top of the Feature Manager, and either select "Edit Material" or pick one of the materials listed. This list can be changed in the Favorites tab in the Materials library. For this part select "Cast Alloy Steel" from the "Steel" library. Click on "Apply" to accept the material and Close the library.

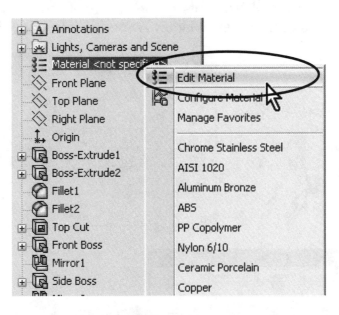

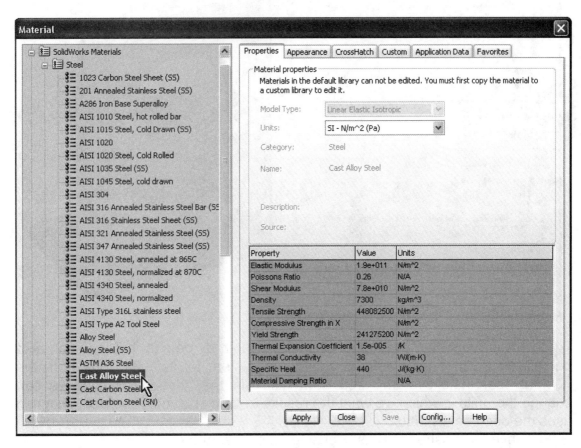

Now the Feature Manager reads "Cast Alloy Steel" instead of "Material". Select **"Mass Properties"** from the "Evaluate" tab in the Command Manager. We'll see the Density (provided by the material selection), Mass (Calculated from the volume and density), Volume, Surface Area and Center of Mass coordinates relative to the origin (also indicated by a magenta triad in the graphics area), Principal Axes of inertia and Moments of inertia about the Center of Mass and the part's origin all listed in a new window, were we can copy the text for later use in reports.

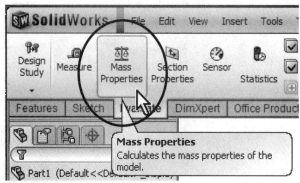

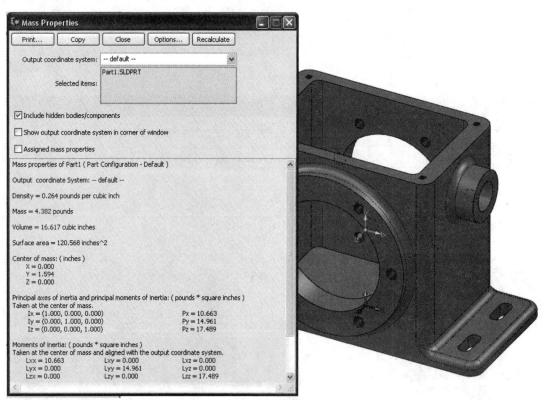

Mass properties are referenced to the origin by default, but they can also be referenced to a user defined coordinate system by selecting one from the "Output coordinate system" drop down list. In the "Options" button we can change the units in which we want the results to be displayed. The default option uses the document's units.

Save the finished part as "Housing" and close the file.

Exercises: Build the following parts using the knowledge acquired in this lesson. Try to use the most efficient method to complete the model.

Exercise 1   Units: inches
Material: AISI 1020 Steel

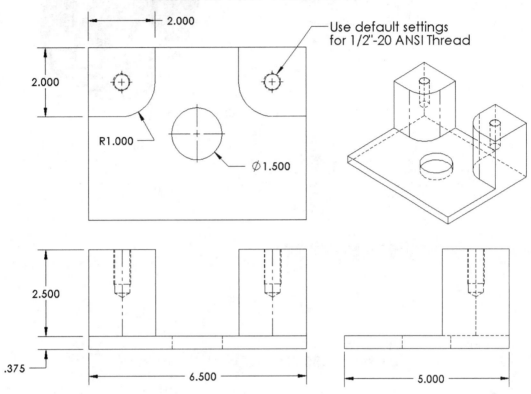

2.000

Use default settings for 1/2"-20 ANSI Thread

2.000

R1.000

Ø1.500

2.500

.375

6.500

5.000

Exercise 2   Units: inches
Material: Al 6061 Alloy

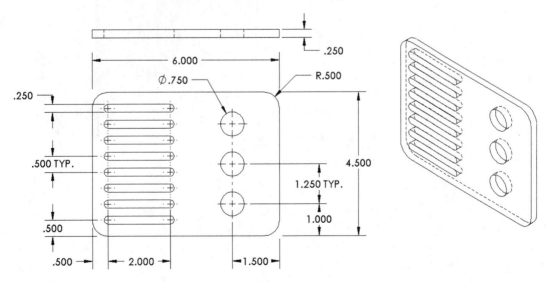

.250

6.000

Ø.750

R.500

.250

.500 TYP.

4.500

1.250 TYP.

1.000

.500

.500

2.000

1.500

Exercise 3   Units: mm
Material: Copper

R2.5 TYP.

50

15

Ø20

Ø115

45

M12x1.5 Tapped Hole
Default depth settings
(4X)

Ø80

Ø150

25

Exercise 4   Units: mm
Material: ABS Plastic

130

25

10

30
TYP.

10

10

R5

20   40

160

M6x1.0 Tapped Hole
Default depth (2X)

40

15

R10

15

40

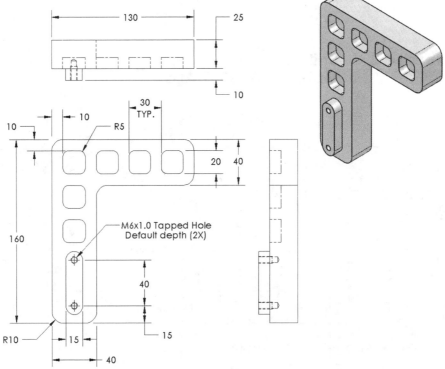

The Side Cover

In making the Side Cover part we will learn the following features and commands: Revolved Feature, Sketch Trim and Extend, and construction geometry. We will also review some of the commands previously learned in the Housing part. The sequence of features we'll follow for the Side Cover is:

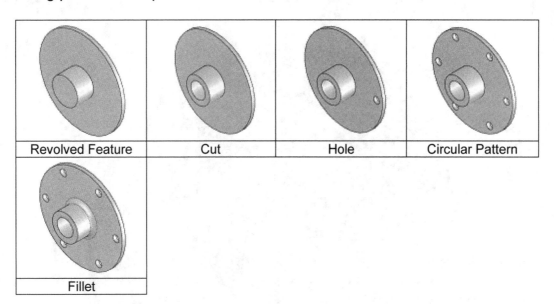

Revolved Feature	Cut	Hole	Circular Pattern

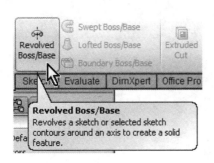

Fillet

67. - Let's make a new part document. Select the "Part" template and click OK.

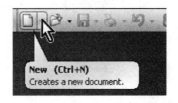

The first feature we'll create is a **Revolved Feature**. As its name implies, it is created by revolving a sketch about an axis. Select the "**Revolved Boss/Base**" icon from the "Features" tab in the Command Manager. When asked to select a plane for the sketch select the "Right Plane". No particular reason to choose this plane, but to have a nice isometric view ☺.

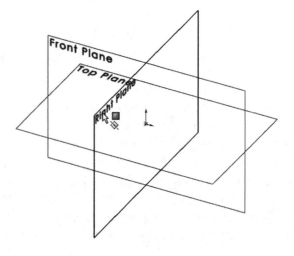

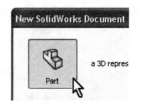

68. - Select the "Rectangle" icon from the Sketch Toolbar and draw the following sketch using two rectangles, starting at the origin and to the left (there will be two lines overlapping in the middle). Don't be too concerned with their size; we'll add the correct dimensions later.

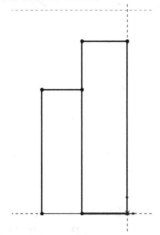

69. - It is a good practice to have single, non-intersecting profiles, and no more than 2 lines sharing an endpoint. It is possible to use a sketch with intersecting lines using a function called "Contours", but it will not be covered in this book.

 As a general rule, it is always a good idea to work with single contour sketches, and advance to other techniques like contours later on.

To 'clean up' the sketch, we will use the "**Trim**" command.

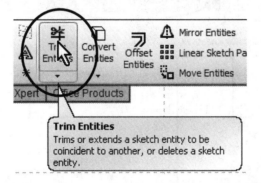

70. - The Trim tool allows us to cut sketch entities using other geometric elements as a trim boundary. After selecting the "**Trim Entities**" icon, select the "**Power Trim**" option from the Property Manager. The Power Trim allows us to click-and-drag *across* the entities that we want to trim. Click and drag the cursor crossing the two lines indicated next. Notice as you cross them they will be trimmed.

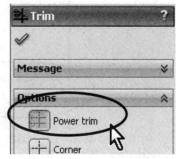

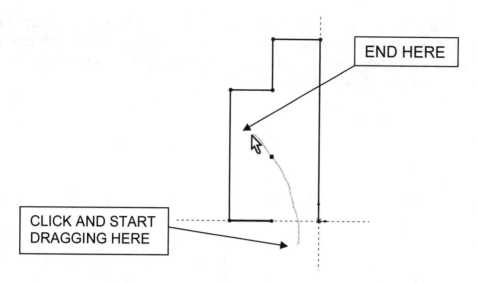

END HERE

CLICK AND START
DRAGGING HERE

71. - The next step is to extend the short line to close the sketch and have a single closed profile. Select the "**Extend Entities**" icon from the drop down menu under "Trim Entities". Click on the short line indicated; a preview will show you how the line will be extended. If the extension does not cross a line, you will not get a preview.

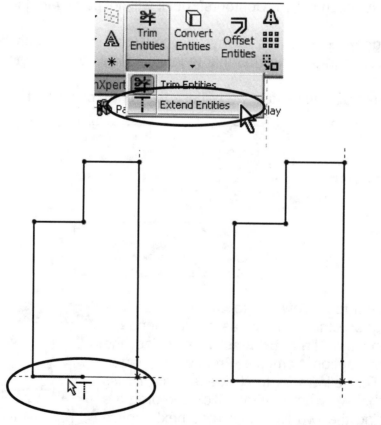

Another way to extend the line is to click-and-drag it's endpoint to the origin.

72. - Add the following dimensions to the sketch using the "Smart Dimension" tool from the "Sketch" tab in the Command Manager or the Mouse Gestures.

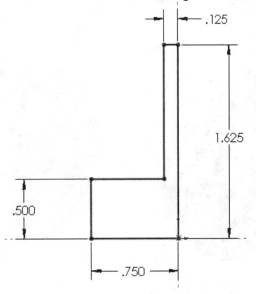

73. - Now that the Sketch is fully defined we'll make the **Revolved Boss/Base**. Select "**Exit Sketch**" or the Revolved Boss/Base icon from the "Features" tab.

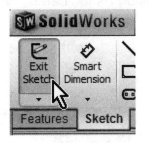

74. - The **Revolve** Property Manager is presented and waiting for us to select a line or centerline to make the revolved feature about it; if the sketch has one centerline, it is selected automatically as a default axis of rotation. Select the line that we extended as the axis of rotation to make the revolved base.

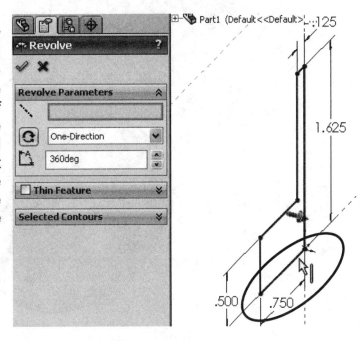

75. - Notice the preview when the line is selected. The default setting for a revolved feature is for 360°. Click OK to complete the revolved feature. Rename the feature "Flange Base".

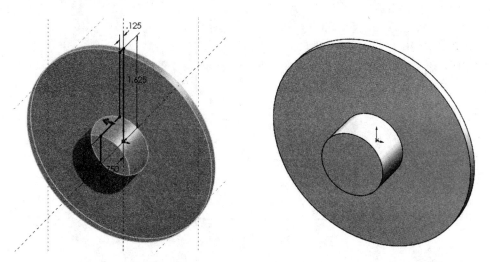

76. - Now switch to a "Front view". We'll make a hole in the center of the cover for a shaft. Create a new sketch in the front most face of the cover (Small round face), or, select the small face and click in the "Sketch" icon from the pop up toolbar. Draw a circle starting at the origin and dimension as indicated. Make a cut using the "Through All" option.

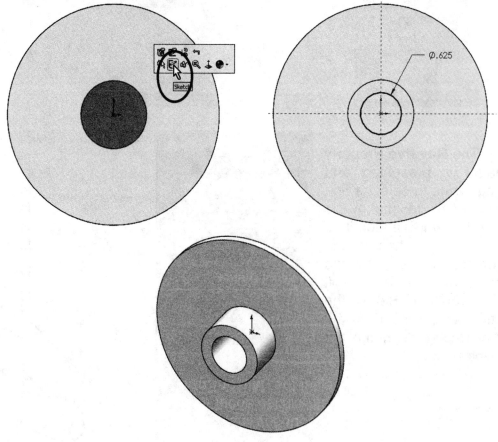

77. - The next step will be to make the first hole for the screws to pass through. We'll make one hole, and then use a Circular Pattern to make the rest as we did in the Housing. In this case we'll use the "Cut Extrude" feature and not the Hole Wizard to show a different approach. To make this hole select the large circular face and create a new sketch. Draw a centerline from the origin to the right, and at the end of the centerline draw a circle. Dimension as shown and make a Cut using the "Through All" option.

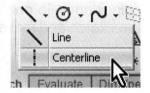

 The centerline is used as reference geometry to locate the center of the circle. Optionally we can add a Horizontal relation between the circle's center and the origin.

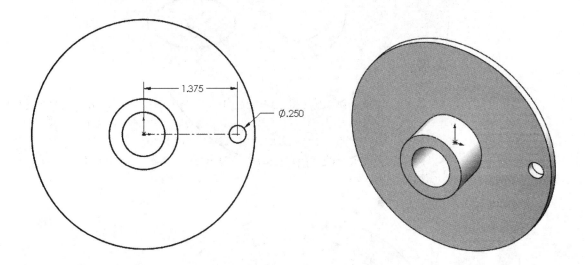

78. - A different way to do this sketch is to draw a circle and then convert it to construction geometry, this way you can dimension the circle's diameter. To convert *any* sketch element to construction geometry, simply select it in the graphics area and activate the "For construction" check box in the element's Property Manager or click the "**Construction Geometry**" icon from the pop up toolbar. Notice the circle is now displayed as construction geometry.

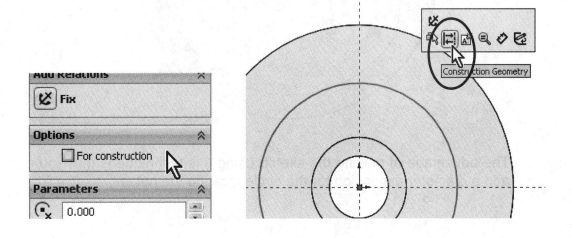

Now that the location circle is construction geometry (also known as *reference geometry*), draw the second circle for the hole making its center coincident with one of the quadrants of the location circle as shown. Dimension the sketch and make the cut.

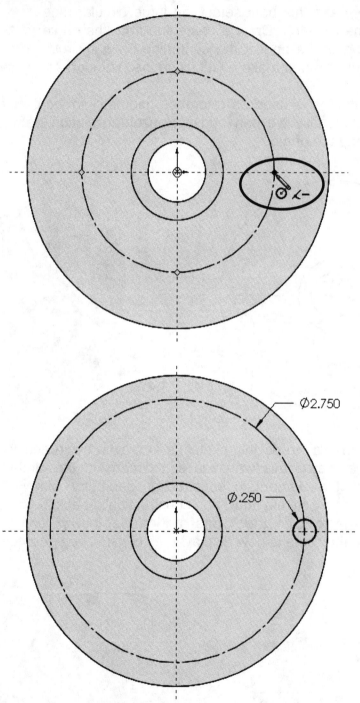

 The advantage of making the sketch using this technique is that you can add a diameter dimension for the circle locating the holes.

79. - To complete the rest of the holes, we'll make a circular pattern. Select the "**Circular Pattern**" icon from the "Features" tab in the Command Manager and select any circular edge for the pattern direction.

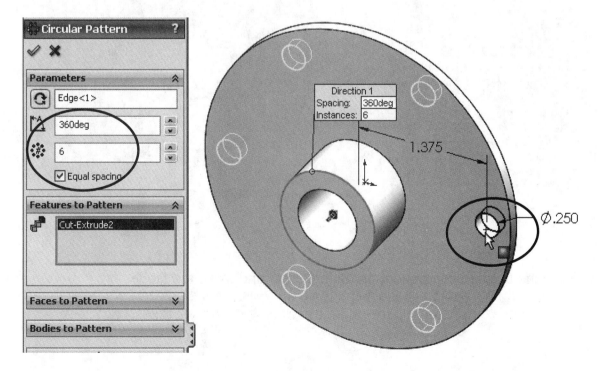

Now click inside the "Features to Pattern" selection box, and if not already selected, select the last cut operation from the fly-out Feature Manager or in the graphics area. Using the "Magnifying Glass" (Shortcut "G") may help. Change the number of copies to 6; remember this count includes the original. Click OK to complete.

80. - For the final step, select the Fillet command and round the edge as shown with a 0.125" fillet radius.

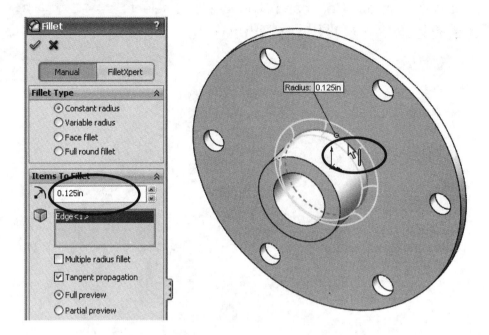

81. - Edit the material for the part, and assign AISI-1020 steel. If it is not in the materials favorite list, select it from the library.

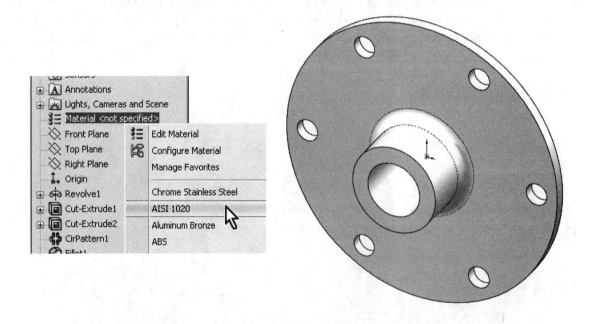

Save the finished component as "Side Cover" and close the file.

Exercises: Build the following parts using the knowledge acquired in this lesson. Try to use the most efficient method to complete the model.

Exercise 5   Units: inch

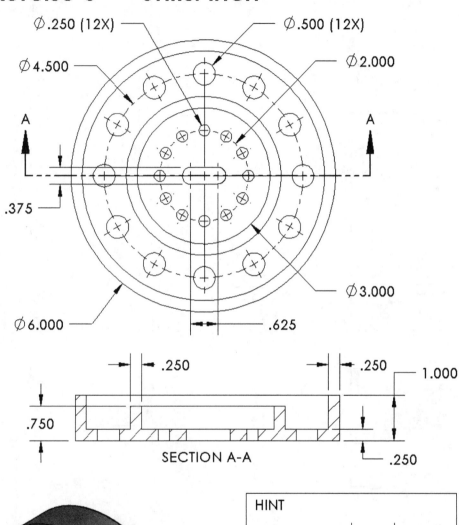

SECTION A-A

HINT

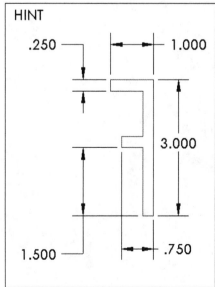

Exercise 6   Units: mm

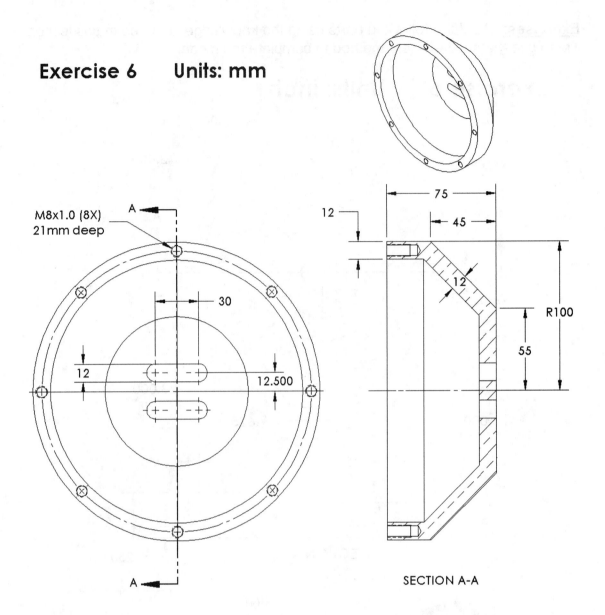

M8x1.0 (8X)
21mm deep

30

12

12.500

75

12

45

12

R100

55

SECTION A-A

The Top Cover

For the Top Cover part we will follow the next sequence of features. In this part we will learn a new feature called Shell, new options for Fillet and new extrude end conditions. Plus, we'll practice some of the previously learned features and options.

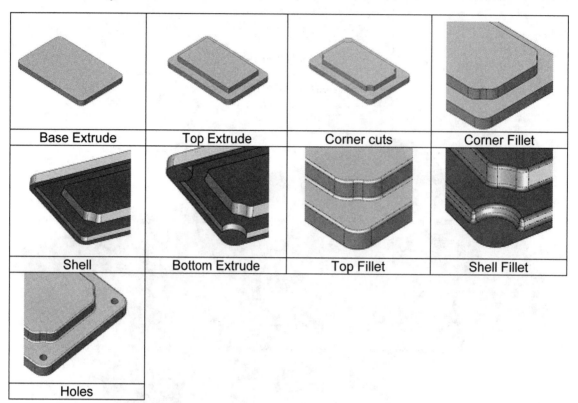

Base Extrude	Top Extrude	Corner cuts	Corner Fillet
Shell	Bottom Extrude	Top Fillet	Shell Fillet
Holes			

82. - We'll start by making a new document selecting the "Part" template from the New Document command, and just like we did with the Housing, we'll create a new sketch on the Top plane. Select the "**Extruded Boss**" icon and then select the "Top Plane".

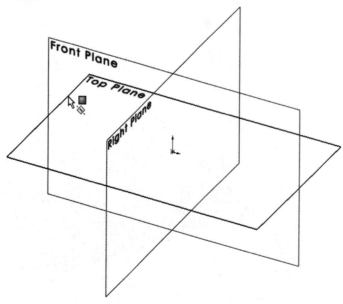

Draw a rectangle using the "**Center Rectangle**" command; first click in the origin to locate the center and then outside to complete the rectangle. Add the dimensions shown with the "**Smart Dimension**" tool and finally round the corners with the "**Sketch Fillet**" command using a 0.25"radius.

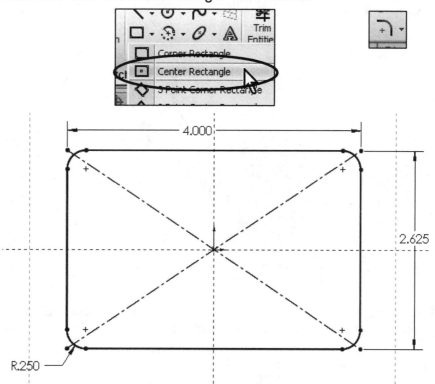

83. - To make the first extrusion, click on "**Exit Sketch**" and set the distance to 0.25". Rename the extrusion "Base".

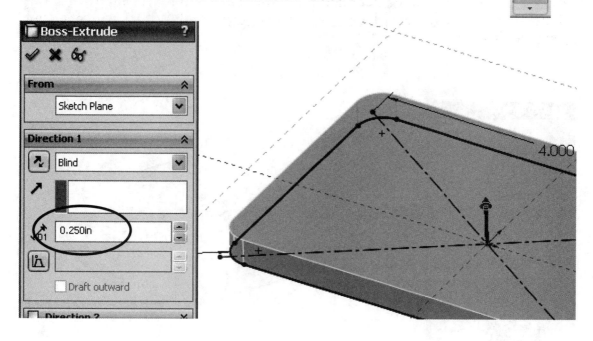

84. - For the second feature, we'll make an extrusion of similar shape to the first one, but smaller. For this feature we will use the "**Offset Entities**" function, this way the sketch will be created automatically by offsetting the edges of the previous feature's face. Click in the top face of the model and select the "Sketch" icon from the pop up toolbar to create a new sketch.

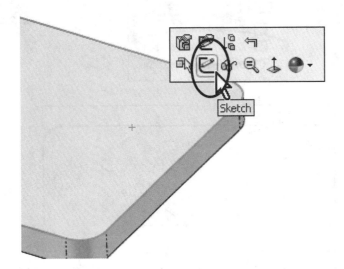

85. - After creating the sketch notice the top face of the first feature is selected (Colored green); while it is selected click the "**Offset Entities**" icon in the "Sketch" tab of the Command Manager. You will immediately see a preview of the offset of the face edges.

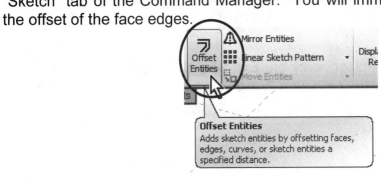

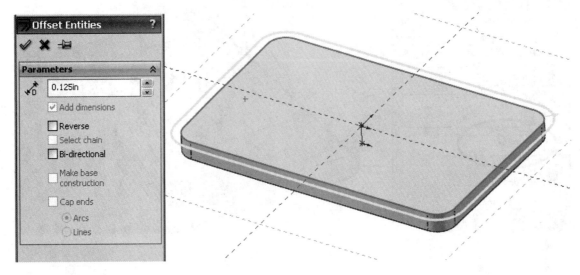

86. - Change the offset value to 0.375" and then click in the "**Reverse**" checkbox to make the offset inside, and not outside. Notice that when we change the direction the preview updates accordingly. Click OK when done.

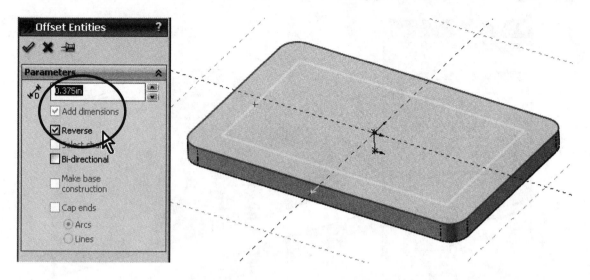

87. - The Sketch is now Fully Defined (all geometry is black) because the sketch geometry is related to the edges of the face and only the offset dimension is added.

 Notice the offset command is powerful enough to eliminate the rounds in the corners if needed.

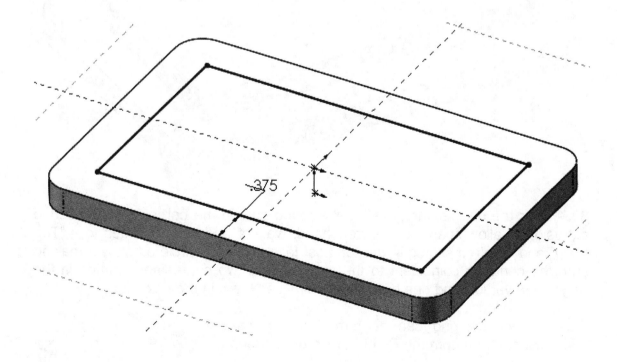

88. - To make the second feature select "**Extruded Boss**" from the "Features" tab in the Command Manager and extrude it 0.25". Rename the feature "Top Boss"

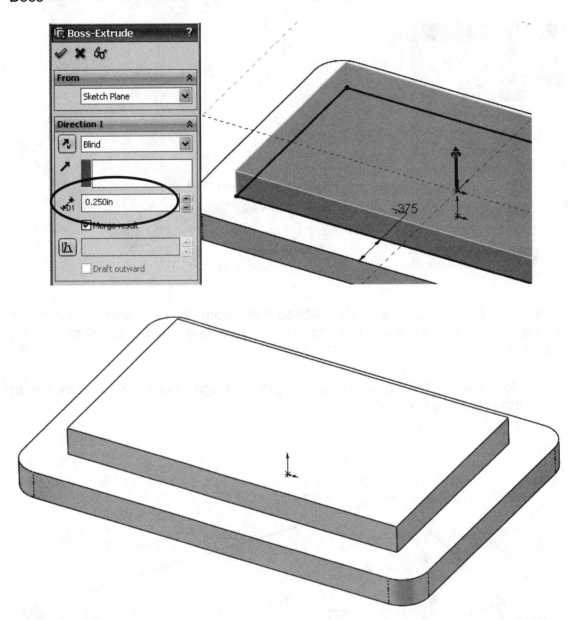

89. - For the next feature, we'll make round cuts in the corners of the second extrusion to allow space for a screw head, washer and tools. Switch to a "Top View" and create a sketch in the topmost face. Make a circle as shown making sure the center is coincident to the corner. Add two centerlines starting in the origin, one vertical and one horizontal. We'll use them in the next step.

 The sketch grid can be turned off for better visualization from the Right-Mouse-button menu.

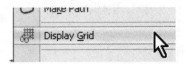

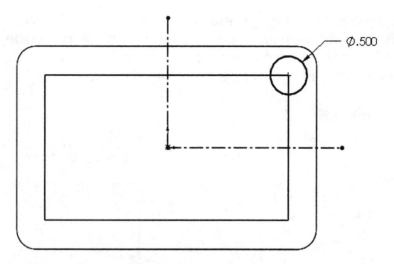

90. - We will now use the "**Mirror Entities**" from the "Sketch" tab in the Command Manager. This tool will help us make an exact copy of any sketch entity, in this case the circle, about any line, edge, or for our example, the vertical centerline; then we'll copy both circles about the horizontal centerline to make a total of four equal circles. Select the "**Mirror Entities**" icon.

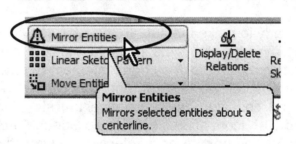

91. - In the "Entities to Mirror" selection box select the circle. Then click inside the "Mirror About" selection box to activate it (it will be highlighted) and select the vertical centerline. Click OK to complete the first sketch mirror.

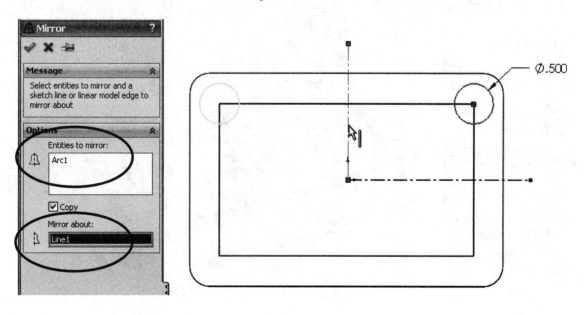

92. - Now repeat the "**Mirror Entities**" command selecting both circles under the "Entities to Mirror" selection box, and the Horizontal centerline in the "Mirror About" selection box. Since the new circles are mirror copies of the original, and the original was fully defined, the sketch is fully defined.

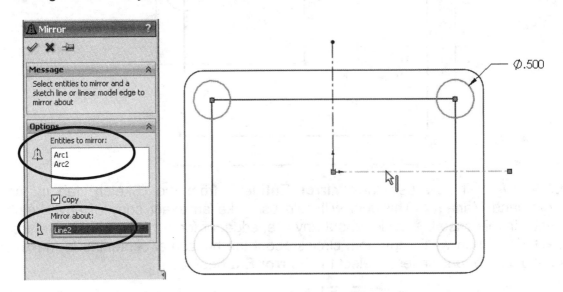

93. - Now we are ready to make the cut. In this step we'll cut all four corners at the same time. SolidWorks allows us to have multiple closed contours in a sketch for one operation as long as they don't intersect or touch at one point. To add intelligence to our model, we'll use an end condition for the cut called **Up to Surface**; with this end condition we can define the stopping face for the cut. Select the "**Extruded Cut**" icon, and select "Up To Surface" from the "Direction 1" options drop down selection box. A new selection box is activated; this is where we'll select the face where we want the cut to stop. Select the face indicated as the end condition and click OK to finish the feature.

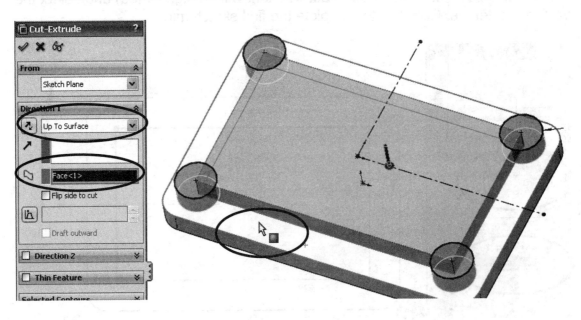

The idea for selecting a face as an end condition is that if the height of the extrusion changes, the cut will still go to the intended depth. This is the design intent and how we keep it by adding intelligence to our model.

Our part should now look like this:

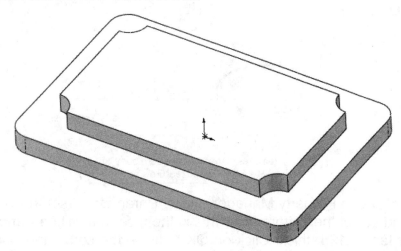

94. - Using "Fillet" from the "Features" tab add a 0.25" radius fillet to the edges indicated to round the corners.

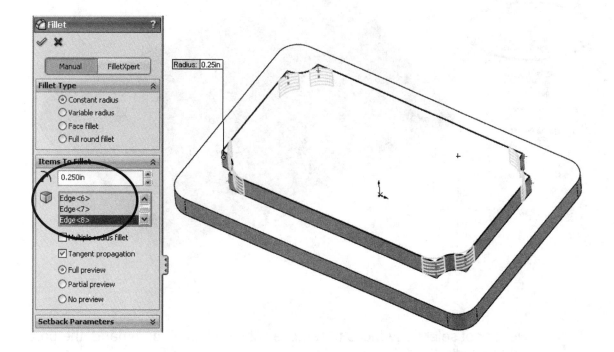

 If selecting the small edges is difficult, try using the "Magnifying Glass" to make selection easier. Notice that there are eight (8) edges to be rounded.

95. - Since this is going to be a cast part, we want to remove some material from the inside, and make its walls a constant thickness (common practice for castings as well as plastic parts). The "**Shell**" command is the best tool for this purpose. The Shell creates a constant thickness part by removing one or more faces from the model. Select the "**Shell**" icon from the "Features" tab in the Command Manager.

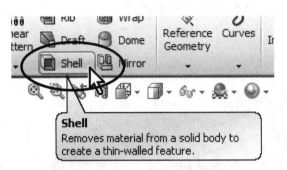

96. - In the Shell's Property Manager under "Parameters" set the wall thickness to 0.125" and select the bottom face; this is the face that will be removed making every other face 0.125" thick. Click on OK to finish the command. Since we only have one shell feature in this part, there is no real need to rename it.

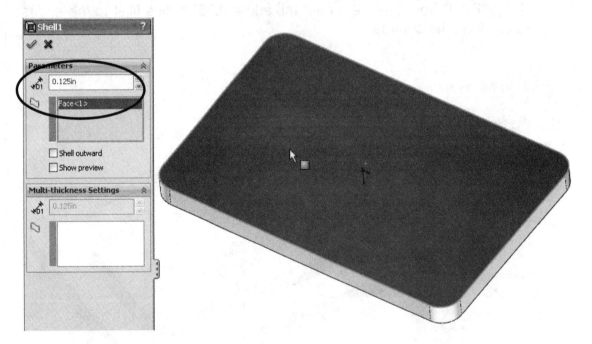

 If we do not select any faces to remove with the "Shell" command, the part would be hollow.

97. - With the finished shell operation the part looks like this, with every face in the part 0.125" thick.

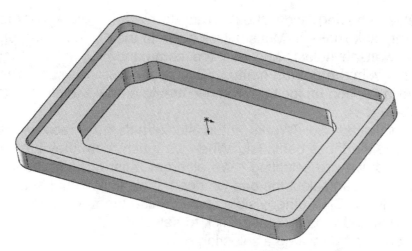

How can we tell the walls are really 0.125"? By using the "**Measure**" tool. It is located under the "Evaluate" tab in the Command Manager. This tool is like a digital measuring tape, where we can select faces, edges, vertices, axes, planes, coordinate systems or sketch entities to measure to and from.

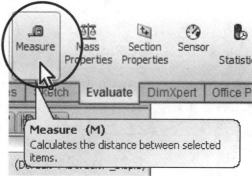

Activate the "**Measure**" tool and select the indicated face and edge, notice the result in the Measure window as well as the tag (Dist .125in) in the graphics area.

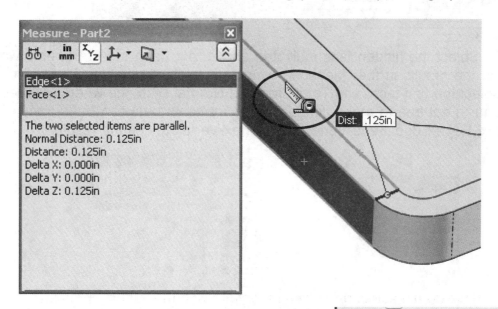

 To measure a different set of entities, click on an empty area of the graphics window, or make a Right-Mouse-Click inside the selection box, and select "**Clear Selections**". This option works with every selection box.

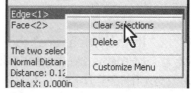

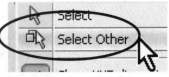

98. - To select a hidden face without having to rotate the component, make a Right-Mouse-Click close to the face that we want to select, and use the command **"Select Other"** from the pop up menu. (This option is also available in the pop up toolbar when we create a new sketch).

When activated, SolidWorks automatically hides the face we clicked on. Now we can see the faces behind it. When we touch them they are highlighted, so we know what we are selecting. We also get a list of faces behind the one removed where we can select the one we need. If we still cannot see the face we need to Right-Mouse-Click other visible faces to hide them as needed. When we see the face we need to Left-Mouse-Click to select it. Every hidden face will be made visible again after making a selection.

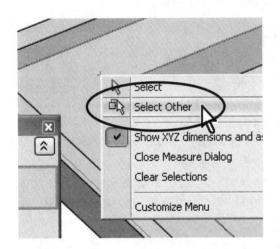
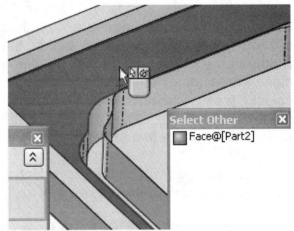

Select the hidden face indicated above, and then select the face we hid. Notice the results in the "Measure" window including Area and Perimeter when we select the first face, and Total area and Normal Distance when we select the second. Feel free to make different measurements between Edge-Face, Edge-Vertex, Edge-Edge, Face-Vertex and Vertex-Vertex to see the results.

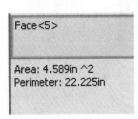
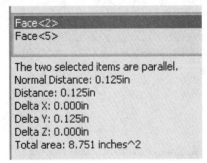

In the "Measure" tool options, we can change the units of measure if needed.

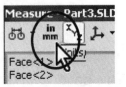

99. - The shell operation left the outside corners thin and now we have to add material to reinforce them by adding an extrusion and have enough support for screws in the corners. Switch to a "Bottom View" (Ctrl+6) and create a sketch on the bottom most face as indicated, be sure to make the circles concentric to the corner fillet edges. Dimension one circle 0.5" diameter, and don't worry too much about the size of the other circles; we'll fix them in the next step.

 Remember to touch the round edges to reveal their centers, and then draw the circle starting in the center to make them concentric.

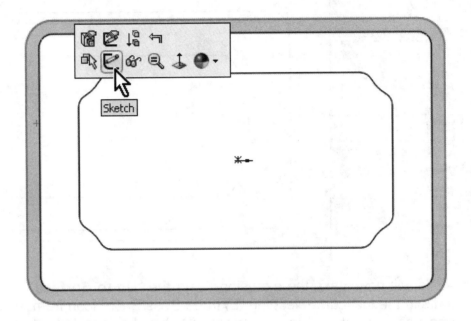

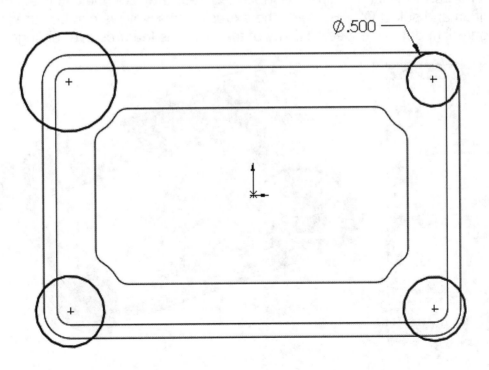

100. - In this case, instead of making the other circles using the "Mirror Entities" command as we did before, we will use sketch relations to make all circles equal. This is just to show a different technique to accomplish the same result. Select the "**Add Relations**" icon from the "Sketch" Tab in the Command Manager or from the right mouse button pop up menu. Select all four circles and under "Add Relations", select **Equal** to make all circles the same size. By dimensioning only one circle, the sketch is now fully defined.

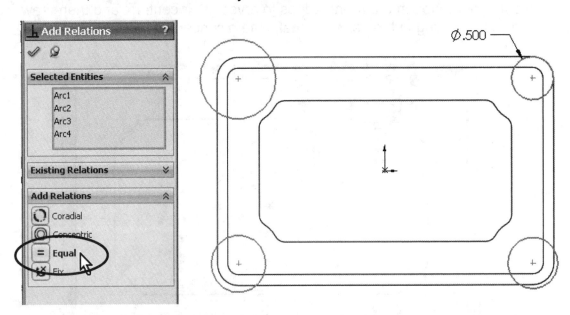

101. - We will now make the extrusion using the "**Up to Surface**" end condition as we did in the last cut. Select the "Extrude Boss/Base" icon from the "Features" tab, and use the option "Up to Surface". Select the face indicated as the end condition and select OK. By doing the extrusion this way we can be sure that our design will update as expected if any of the previous features ever change.

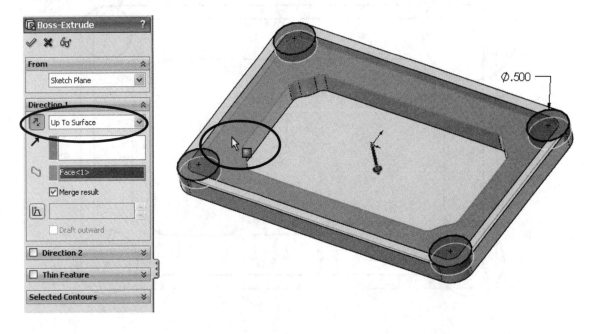

Our part should now look like this:

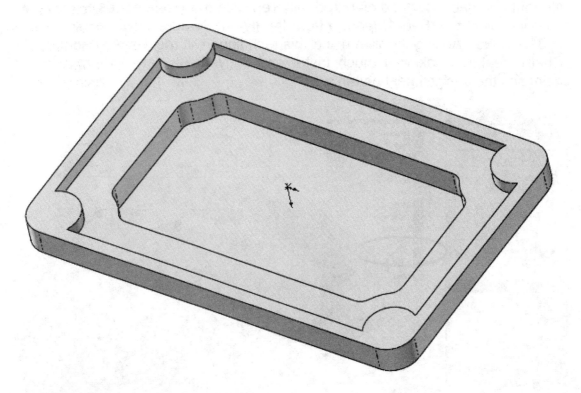

102. - Add a 0.031" radius fillet using the "**Fillet**" command from the "Features" tab. Select the two faces on top of the cover as indicated. Notice all the edges on the top side of the part are rounded with only two selections, maintaining our design intent and making our job easier at the same time.

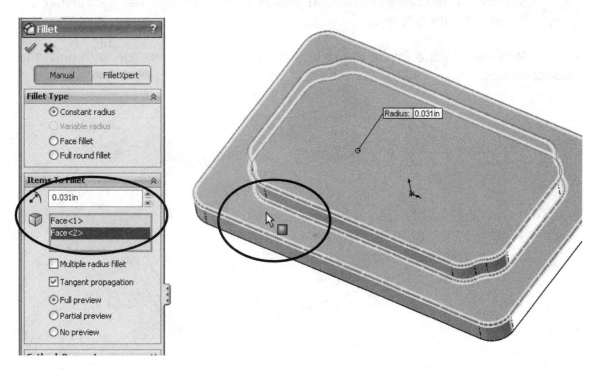

103. - To add fillets to all the inside edges of the part we will use a slightly different approach. Instead of individually selecting the inside edges or faces, we will only select the "Shell1" feature from the fly-out Feature Manager, and add a 0.031" radius. Adding the fillet using this technique will round every edge of the "Shell1" feature, making it much faster and convenient, not to mention that it maintains the design intent better.

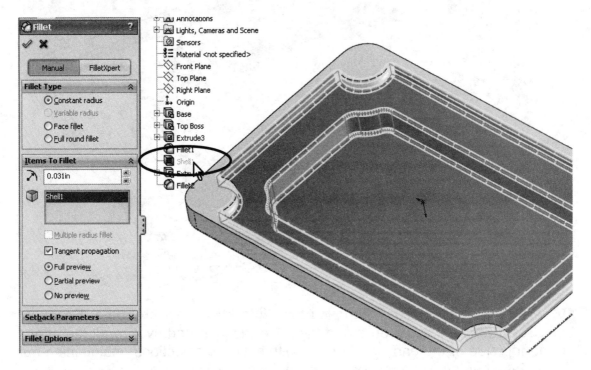

 Selecting a feature to fillet as in this example can only be done using the fly-out Feature Manager.

Now every edge inside the part is rounded in one operation with only one selection.

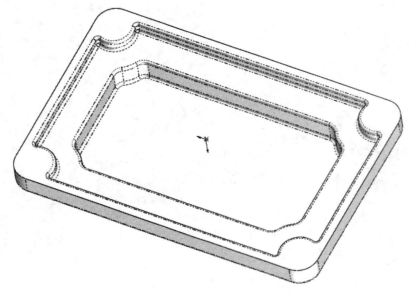

104. - We are now ready to make four clearance holes for the #6-32 screws on the top cover. We'll use the "**Hole Wizard**" feature from the "Features" tab. Switch to a Top View for visibility and select the "Hole Wizard" icon.

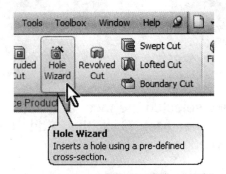

105. - In the first step of the "**Hole Wizard**", select the "Hole" specification icon. From the "Type" drop down list select "**Screw Clearances**". From the "Size" selection list pick "# 6" and from "End Condition" select "Through All". This will create a hole big enough for a #6 size screw to pass freely through it.

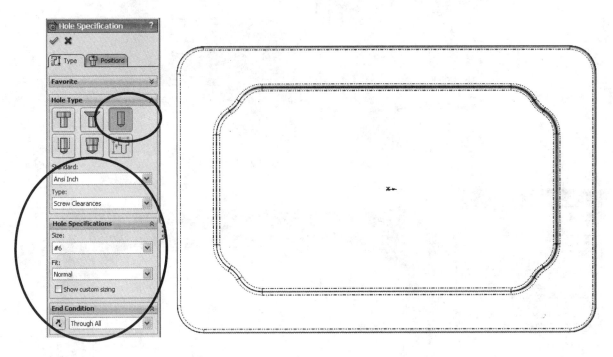

106. - For the second step click in the "Positions" tab and select the face indicated to add the hole locations sketch. Notice we are adding a sketch point where we selected the face. Now we need to add three more points for the rest of the holes. Now that the Point Tool is active touch the round corner edges to reveal their centers as we did in the housing part, and click in their centers to draw the three extra points.

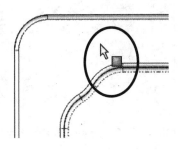

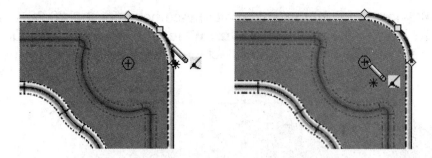

For the first point where we selected the face, add a geometric relation to make it concentric to the fourth corner fillet. Click the "Add Relations" icon, select the sketch point and the round edge.

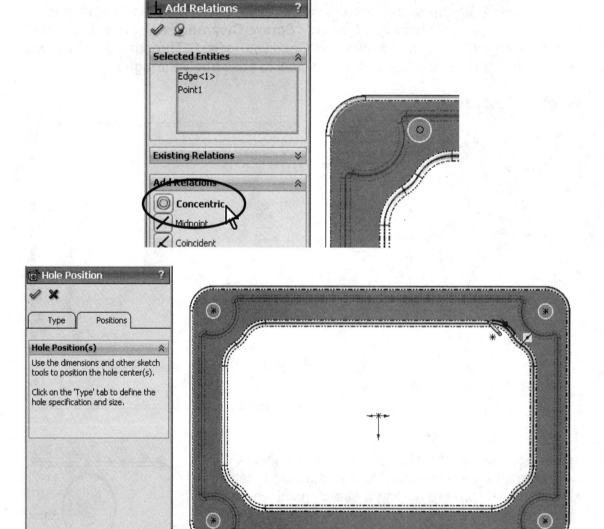

Click OK to complete the Hole Wizard.

107. - Change the part's material to "Cast Alloy Steel" as a final step. Save the part as "Top Cover" and close the file.

Exercises: Build the following parts using the knowledge acquired in this lesson. Try to use the most efficient method to complete the model.

Exercise 7   Units: inch

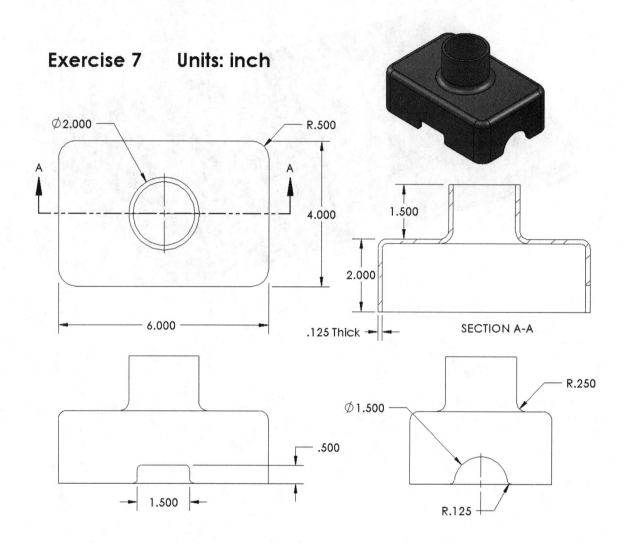

SECTION A-A

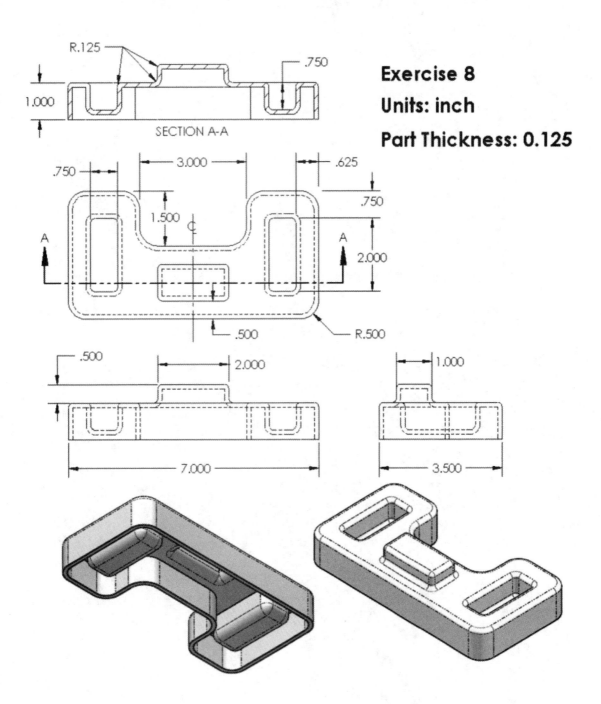

R.125

.750

SECTION A-A

1.000

Exercise 8

Units: inch

Part Thickness: 0.125

.750

3.000

.625

.750

1.500

A

A

2.000

.500

R.500

.500

2.000

1.000

7.000

3.500

Notes:

The Offset Shaft

For the Offset Shaft we'll follow the next sequence of operations. In this part we only need a few features. We'll learn how to make polygons in the sketch, a new option for the Cut Extrude feature, auxiliary planes and a Revolved cut.

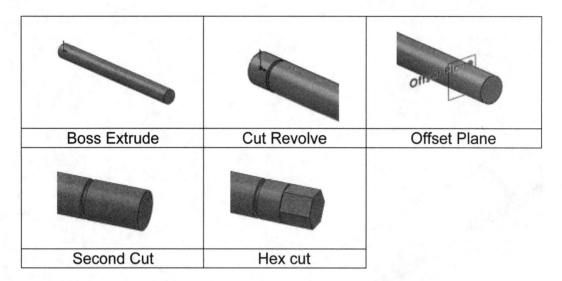

Boss Extrude	Cut Revolve	Offset Plane
Second Cut	Hex cut	

108. - Start by making a new document selecting the Part template. For the first feature in this part create a sketch in the "Right Plane". A different way to create a sketch is to select the "Right Plane" in the Feature Manager and from the pop up toolbar select "Sketch".

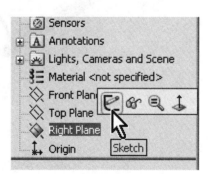

The view will be oriented to the "Right View" automatically. Draw a circle starting at the origin and dimension it 0.600" using the "Smart Dimension" tool. Since this shaft will need to meet certain tolerances for assembly, we will give a tolerance to the diameter.

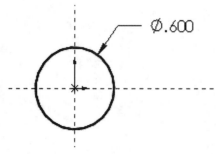

109. - To add (or change) the tolerance of the shaft's diameter, select the dimension in the graphics area. Notice the dimension's properties are displayed in the property manager. This is where we can change the tolerance type. For this shaft select from the "**Tolerance/Precision**" options box "Bilateral" and add +0.000"/-0.005"

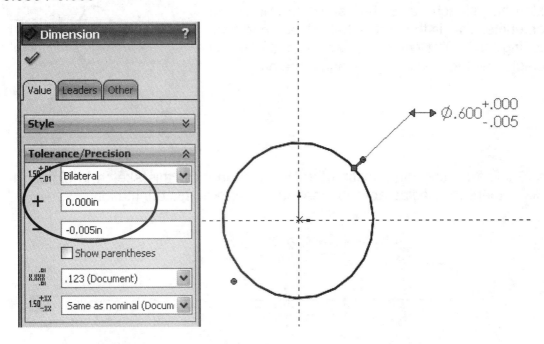

110. - When finished with the tolerance, extrude the shaft 6.5" using "Extruded Boss/Base" from the Features tab.

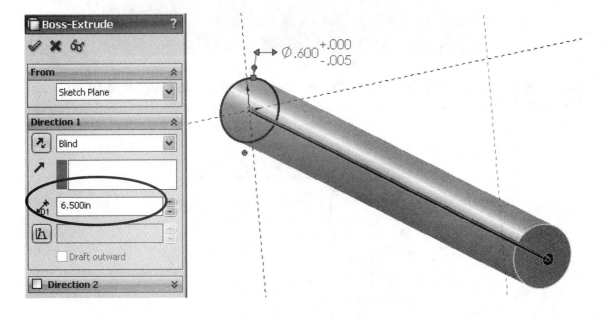

111. - For the second feature we'll make a "**Revolved Cut**". As its name implies, we'll remove material from the part similar to a turning (Lathe) operation. Switch to a "Front View" and select the "Front Plane" from the Feature Manager. From the pop up menu select the "**Sketch**" icon as before to create a new sketch on it. Draw the following sketch and be sure to add the centerline. This is the profile that will be used as a "cutting tool". The centerline will be automatically selected as the axis of revolution for the cut.

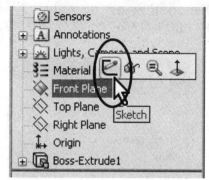

 The reason for selecting the "Front Plane" for this sketch is simply that there are no flat faces to create the sketch in this orientation.

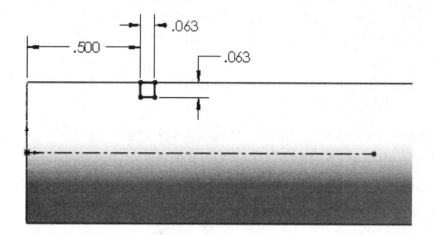

112. - Now that we are finished with the sketch, select the "**Revolved Cut**" icon from the Features toolbar.

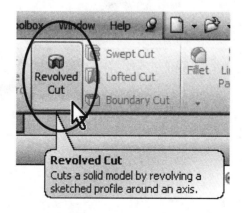

113. - Since we only have one centerline in the sketch, SolidWorks automatically selects it as the axis to make the revolved cut about it. By default a revolved cut is 360 degrees. Rotate the view to visualize the preview. Feel free to use the value spin box to change the number of degrees to cut, this will illustrate the effect of a revolve cut very clearly. Click OK to complete the feature and rename it "Groove".

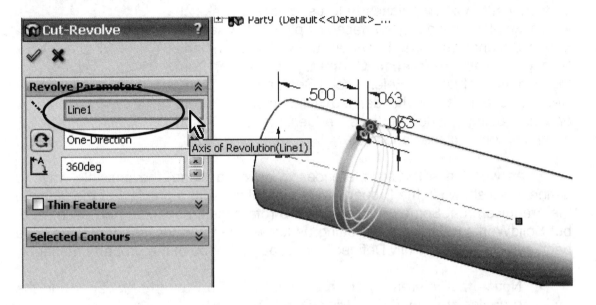

 If there is more than one centerline in the sketch, we will be asked to select one to use as the axis of revolution.

The part should now look like this.

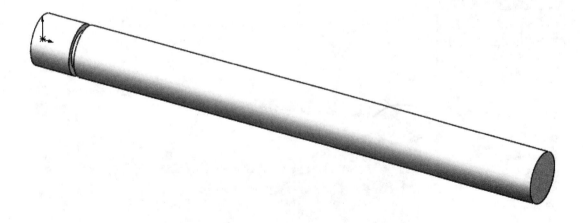

114. - For the next feature, we need to make a cut exactly like the one we just did, but in the right side of the shaft. We could have done it with the Revolved Cut feature before, but we'll show a different way to make it and learn additional functionality at the same time.

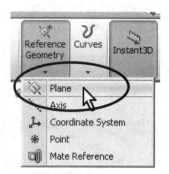

For this feature, we'll create an auxiliary plane for a new sketch. Auxiliary Planes help us to locate a sketch where we don't have any flat faces or planes to use. To create a plane, from the "Features" tab in the Command Manager select "**Reference Geometry, Plane**" or from the menu "Insert, Reference Geometry, Plane". Reference planes can be defined using many different options, using model vertices, edges, faces, sketch geometry, other planes and axes.

As we start selecting references to define a plane, the possible ways to define it are shown in the Property Manager, along with a preview of the plane in the graphics area. Some will use 1, 2 or 3 references, depending on each case, but SolidWorks will help us by letting us know when the necessary options have been selected with a "Fully Defined" message at the top of the property manager.

 Notice that references are color coded in the Property Manager and the graphics area, by default Pink, Purple and Cyan.

Here are some of the most common ways to define a plane:

Through Lines/Points. - Select any 3 non-collinear vertices, or one straight edge and one non-collinear vertex. Two or three references are needed.

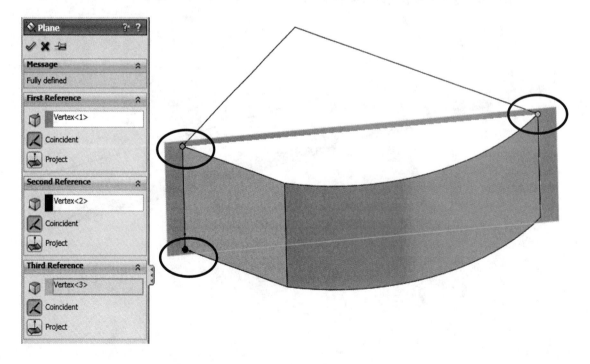

Parallel Plane at Point. – Select an existing Plane or flat face and a point. Note that when we select a Plane or flat face we get additional options like Parallel, Perpendicular, Coincident, at angle or distance. Two references are needed.

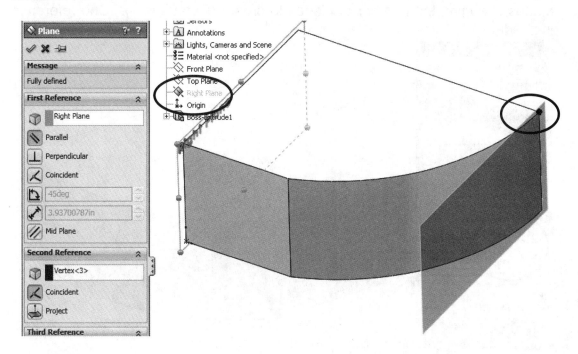

At an angle. – Select a Plane or flat face and an Edge and enter the angle. The Plane's direction can be reversed using the "Flip" checkbox. In this case the face in the back was selected and the Edge indicated. The Edge acts as a "hinge" to the new plane. Change the angle using the value spin box to see the effect. Two references are needed.

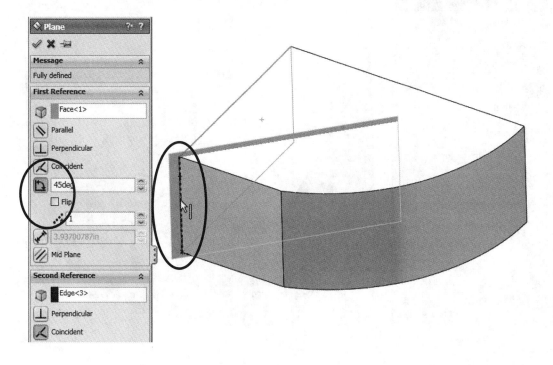

Offset Distance. – Select a Plane or flat face, and define the distance from it to create the new Plane. "Flip" can be used to change the side that the new plane has created. In this case the face in the back was selected and the new plane created 2 inches to the right. We can optionally create multiple parallel planes changing the number of planes to create (Default is 1). One reference is needed.

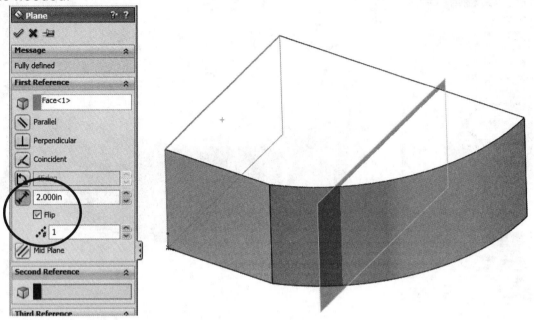

Normal to Curve. – The plane is created by selecting an edge and a vertex of the edge. The plane created is perpendicular to the curve at the vertex. In this case we selected the edge and the vertex indicated. Two references are needed.

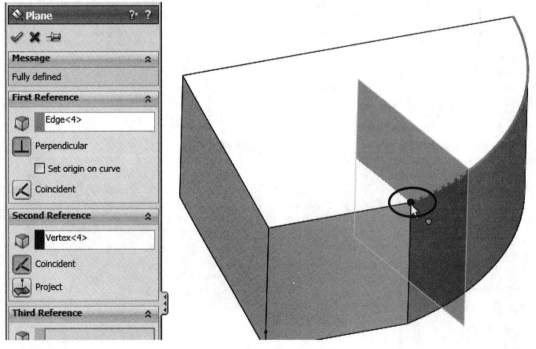

On Surface. – The plane is created selecting a surface (any surface) and a vertex (or sketch point/endpoint) on the surface. In this case the face is curved and the resulting plane is tangent to the face at the vertex. Two references are needed.

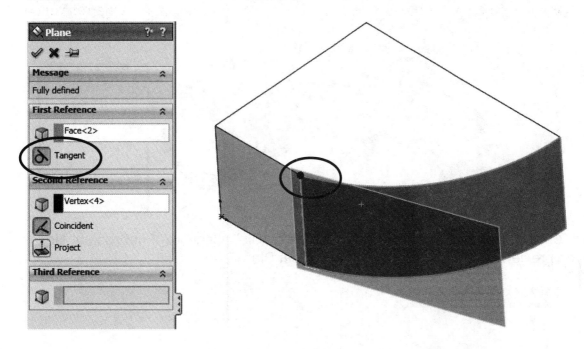

For auxiliary plane creation we can use for reference existing Planes, Faces, Vertices, sketch elements, Axes and Temporary Axes, the Origin, etc. To make sketch elements visible, expand a feature in the Feature Manager, select the sketch that we want to make visible, and from the pop up toolbar select the Show/Hide icon. If the sketch is hidden, it will be made visible and vice versa.

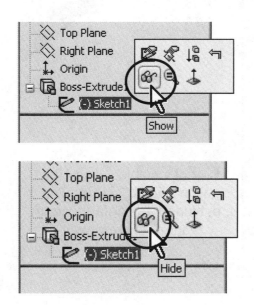

115. - Back to our part, we'll create an Auxiliary Plane parallel to a face of the groove. Select "**Auxiliary Geometry, Plane**" from the Features tab. We'll make it a set distance from the right face of the groove. Using the "**Select Other**" function (Right Mouse button menu or pop-up icon), select the indicated face below.

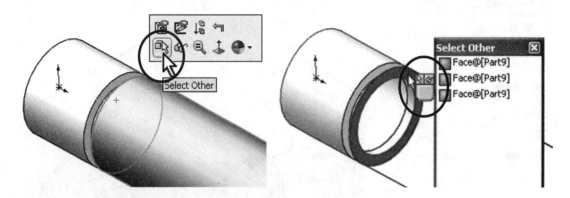

116. - Set the distance to 5" and turn on the "Flip" checkbox if necessary. Notice the preview. Rename the new plane "Offset Plane".

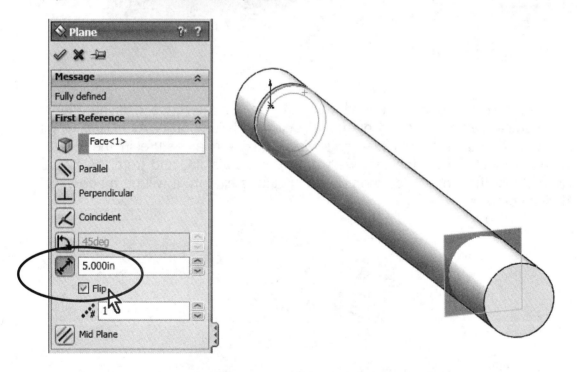

If the new plane is not visible, use the "**Hide/Show Items**" tool to see it. Front, Top and Right Planes are hidden by default; you can hide and show any plane the same way as a sketch.

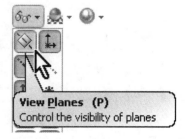

117. - Switch to an "Isometric View" (for visibility) and select the "Offset Plane" in the graphics area; from the pop up toolbar click on "Sketch" to create a new sketch on it.

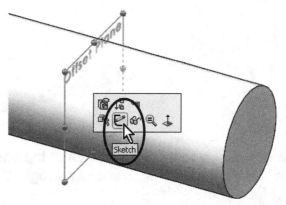

The "**Convert Entities**" command is used to project existing geometric entities onto a sketch, such as model edges and other sketches and convert them to new sketch entities at the same time. Click on "**Convert Entities**" from the Sketch tab in the Command Manager, select the two edges indicated from the "Groove" feature and click on OK.

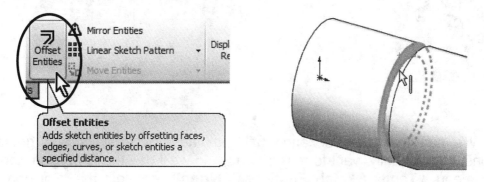

The two edges are projected onto the sketch plane and are automatically fully defined. They are an exact copy of the edges they came from, so if the original groove changes, the converted entities will also. Make an "Extruded Cut" feature 0.063" deep, going to the right. Rename the feature "Offset Groove".

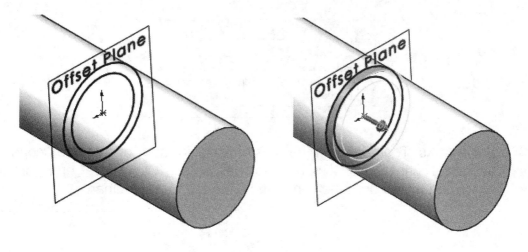

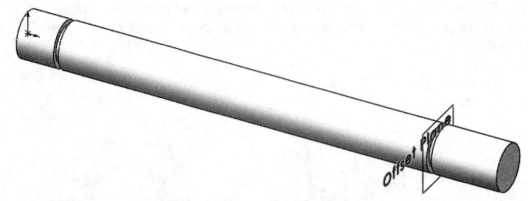

118. - For the last feature, we'll add a hexagonal cut at the right end of the shaft. Hide the "Offset Plane" for easier visibility. Switch to a Right View, and insert a sketch in the rightmost face of the shaft.

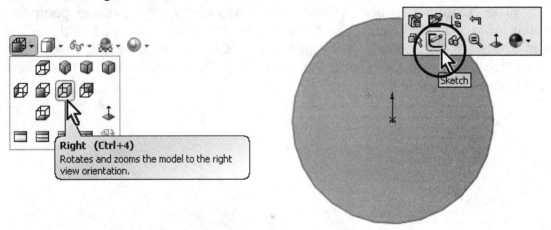

119. - We can make the hexagon utilizing lines, dimensions and geometric relations, but we really want to make it easy, so we'll use the **"Polygon"** tool. Go to the menu, **"Tools, Sketch Entities, Polygon"** or select the "Polygon" icon from the "Sketch" tab in the Command Manager.

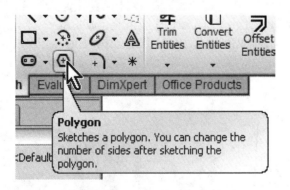

 SolidWorks Toolbars can be customized to add or remove icons as needed. Right-Mouse-Click on a toolbar, select "Customize" and from the "Commands" tab drag the commands needed to the toolbar.

120. - After we select the "Polygon" tool, we are presented with the options in the Property Manager. This tool helps us to create a polygon by making it either inscribed or circumscribed to a circle. For this exercise we'll select the "Circumscribed circle" with 6 sides from the "Parameters" options. Don't worry too much about the rest of the options, as we'll define the hexagon using two more geometric relations.

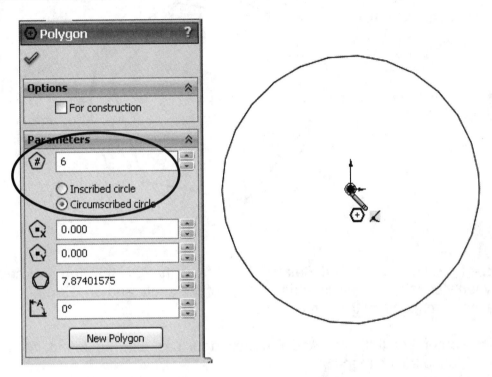

121. - Since we are defining the polygon with a circumscribed circle, we have to draw a circle. Start at the center of the shaft as shown; notice that we immediately get a preview of the hexagon, its radius and angle of rotation.

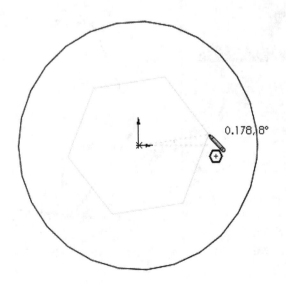

122. - Draw the circle a little smaller (or larger) than the shaft, the idea is to make the circumscribed construction circle the same size as the shaft using a geometric relation. Hit "Esc" or OK to finish the polygon tool when done.

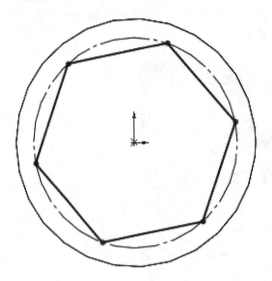

123. - Now select the "**Add Relation**" tool from the "Sketch" tab and select the polygon's construction circle and the edge of the shaft; add an "**Equal**" geometric relation to make them the same size.

 Remember that the "Add Relation" tool can also be found in the Right Mouse button menu.

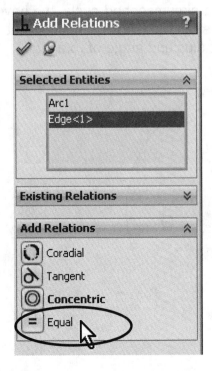

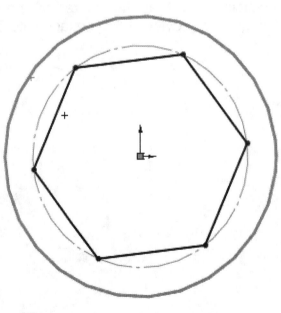

124. - Now select one line of the hexagon, and add a "Horizontal" relation to fully define the sketch.

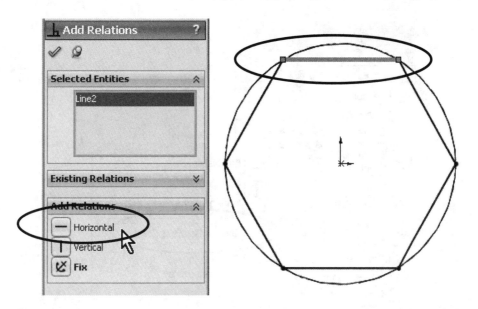

125. - Now we are ready to make the cut. For this operation we'll use a little used but very powerful option in SolidWorks. Select the "**Extruded Cut**" icon from the "Features" tab as before, but now activate the checkbox "**Flip side to cut**". This option will make the cut *outside* of the sketch, not inside. Notice the arrow indicating which side of the sketch will be used to cut. Make the cut 0.5" deep. Rename the feature "Hex Cut".

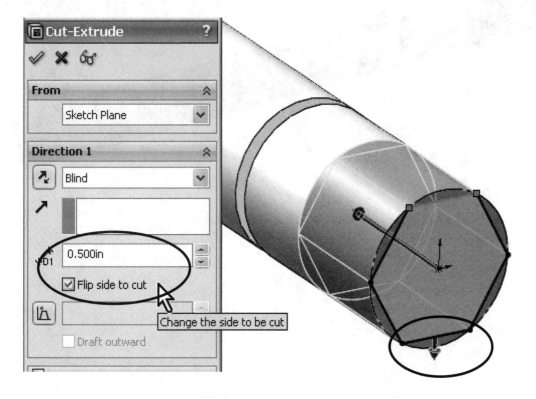

126. - Edit the material for this shaft and select "Chrome Stainless Steel" from the materials library (or the Favorites list if available).

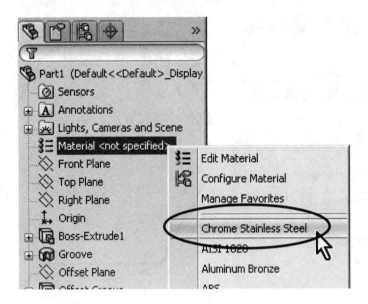

Save the part as "Offset Shaft" and close the file.

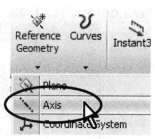

Auxiliary geometry includes **Axes** and user defined **Coordinate Systems.** Both of these can be created from the "Reference Geometry" icon in the "Features" tab in the Command Manager. Reference geometry can be used for a number of reasons, including locating features and components, as reference, or to use as part of a feature (an axis can be used to define the direction for a circular pattern).

Axis can be made using many options, including:

One Line/Edge/Axis.- Any linear edge, sketch line or axis. Every cylindrical and conical face has an axis (temporary Axis) running through it. To reveal it use the menu "**View, Temporary Axis**"

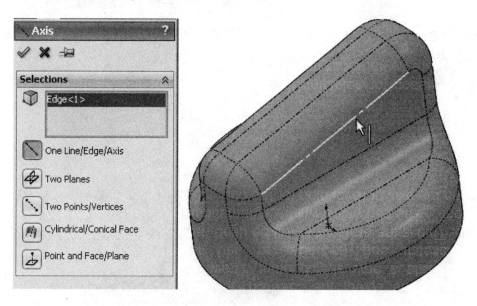

Two Planes. - An axis can be created at the intersection of any two planes.

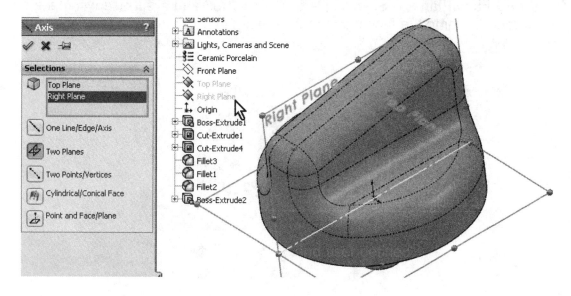

Two points/Vertices. - Using any two vertices and/or sketch points/endpoints.

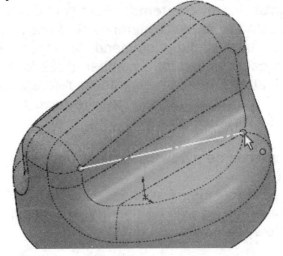

Cylindrical/Conical Face. Selecting any cylindrical or conical face will make an axis using the face's temporary axis.

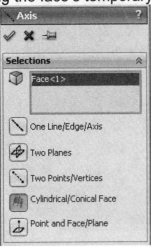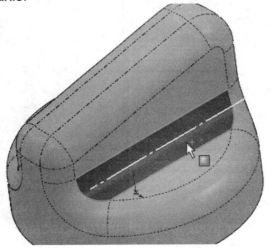

Point and Face/Plane. – Selecting a point or vertex and a surface will make an axis perpendicular to the face/plane that passes through the point/vertex.

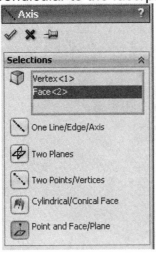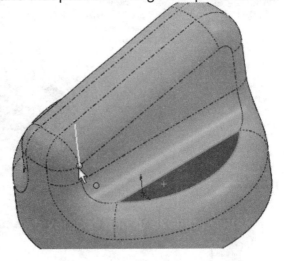

Coordinate Systems can be created to calculate a component's center of mass referenced to a specific location. Another instance when a Coordinate System is helpful is when we have to export parts for manufacturing on computer controlled machines using Computer Assisted Manufacturing (CAM). More often than not the origin the designer used for the design may not be the best location to program the machining equipment.

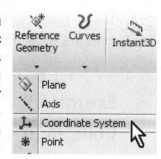

To create a Coordinate System we need to select a vertex or point, and define two axis directions (X, Y or Z) using linear edges, axes or sketch lines, and the third direction will be defined automatically. Notice that an axis direction can be reversed using the "Reverse direction" icon to the left of each selection box.

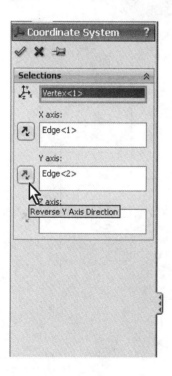

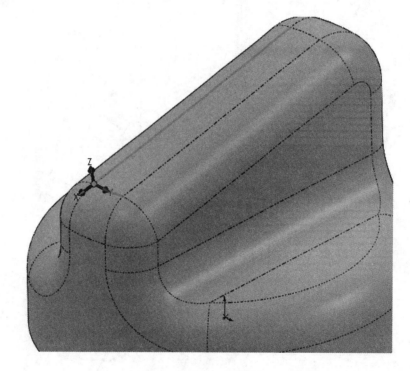

Exercises: Build the following parts using the knowledge acquired in this lesson. Try to use the most efficient method to complete the model.

Exercise 9

Units: inch

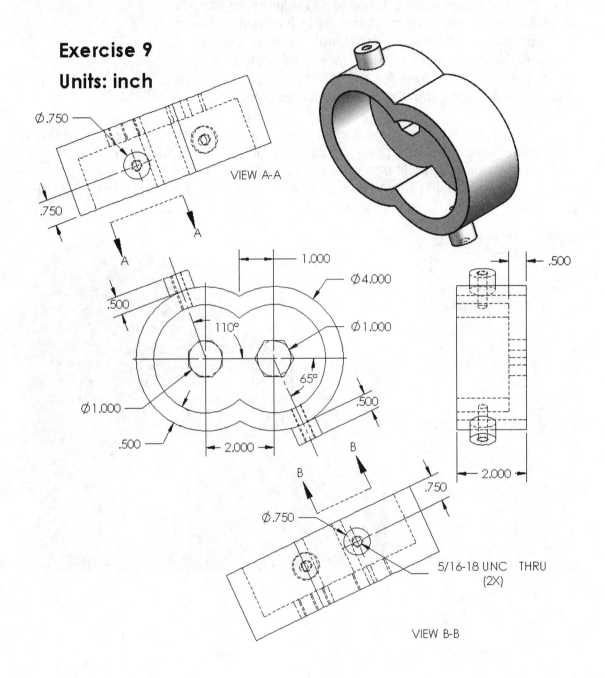

VIEW A-A

VIEW B-B

5/16-18 UNC THRU
(2X)

 Draw a sketch with centerlines, exit the sketch and use it as a layout to make the auxiliary planes.

Exercise 10   Units: inch

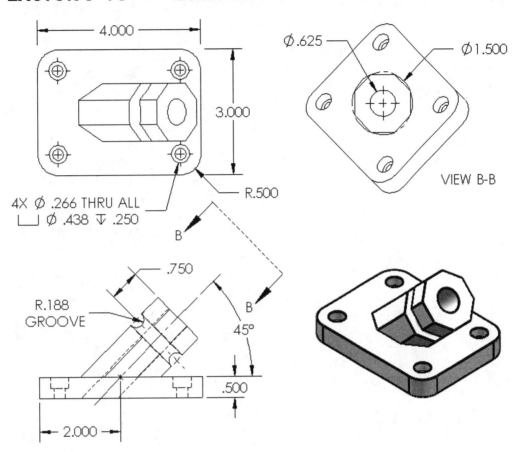

4.000

3.000

R.500

⌀.625

⌀1.500

VIEW B-B

4X ⌀ .266 THRU ALL
⌴ ⌀ .438 ▽ .250

B

.750

R.188
GROOVE

45°

B

.500

2.000

Notes:

The Worm Gear

For the Worm Gear we will make a simplified version of a gear without teeth. The intent of this book is not to go into gear design, but rather to help the user understand and learn how to use basic SolidWorks functionality. With this part we'll learn a new extrusion (or cut) end condition called Mid Plane, how to chamfer the edges of a model, a special dimensioning technique to add a diameter dimension to a sketch used for a revolved feature, and how to add a dimension to a circle's edge. We'll also practice previously learned commands. For the Worm Gear we will follow the next sequence of features.

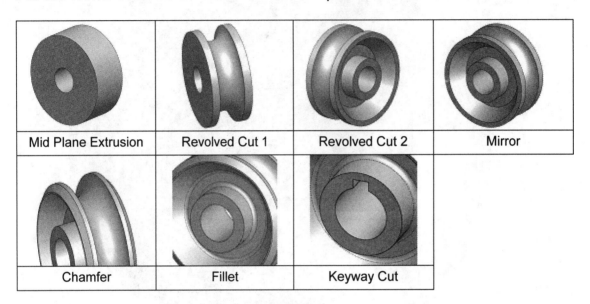

Mid Plane Extrusion	Revolved Cut 1	Revolved Cut 2	Mirror
Chamfer	Fillet	Keyway Cut	

127. - Start by making a new part document and create the following sketch in the Front Plane.

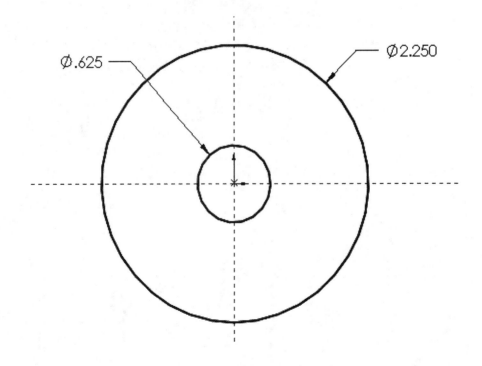

$\emptyset.625$

$\emptyset 2.250$

128. - In this case we want the part to be symmetrical about the Front Plane. To achieve this, we'll make an "Extruded Boss/Base" using the "**Mid Plane**" end condition; this condition extrudes half of the distance in one direction and half in the second direction. Change the end condition to "**Mid Plane**" and extrude it 1" (The result will be 0.5" going to the front and 0.5" going to the back). Rename the feature "Base".

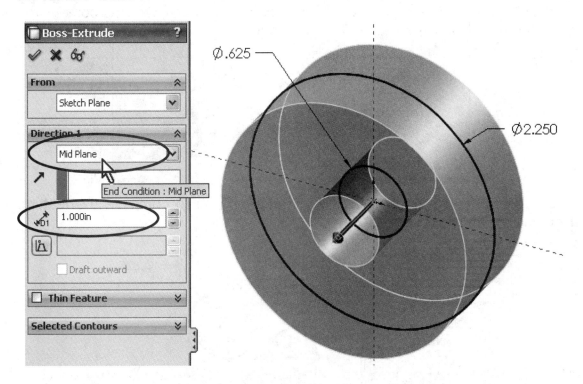

129. - For the second feature we will use a "**Revolved Cut**" to make the slot around the outside perimeter. Switch to a "Right View" and create a sketch in the Right Plane. Select the Right Plane from the Feature Manager, and click in the Sketch icon from the pop up toolbar as shown. Make the center of the circle at the Midpoint of the cylinder's top edge and <u>don't forget the centerline</u>.

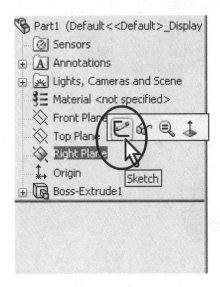
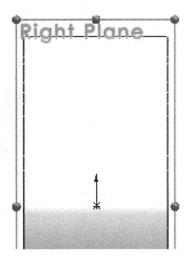

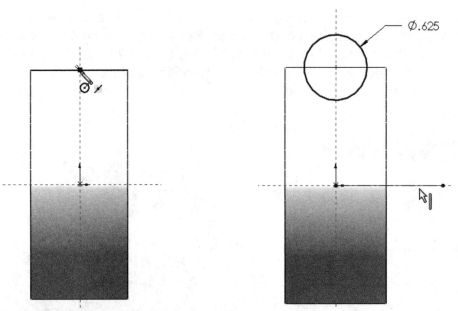

130. - Select "Revolved Cut" from the "Features" tab to complete the feature. Rename the feature "Groove".

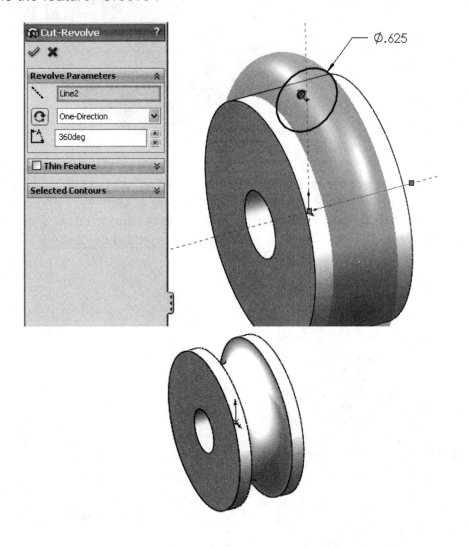

131. - Make a second revolved cut to remove material from one side. Switch to a "Right View", and create a new Sketch in the "Right Plane" as before using the following dimensions. Remember it is a closed Sketch (Four lines, don't forget the vertical line in the right side). Turn off the Grid and change to "Wireframe" view mode for clarity.

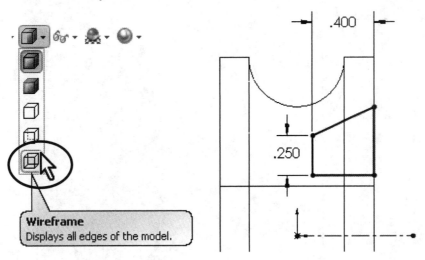

132. - We'll now use a new dimensioning technique to add diameter dimensions for Revolved Features, this way we will have a diameter dimension in the revolved feature when finished. Select the "Smart Dimension" tool, and add a dimension <u>from the Centerline</u> (Not an endpoint!) to the top endpoint of the sketch. Before locating the dimension, cross the centerline and notice how the dimension value doubles. Locate the dimension, and change it to 2". Repeat and add a 1" diameter dimension from the centerline to the horizontal line.

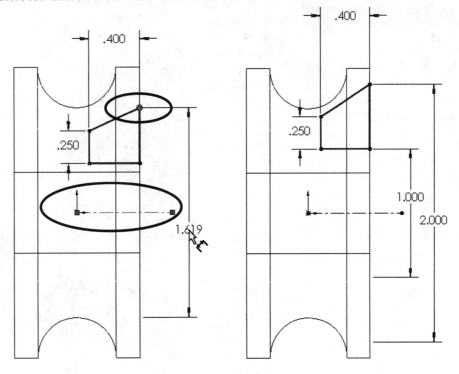

133. - Make a "**Revolved Cut**" about the centerline to complete the feature. Notice the diameter dimensions; their purpose is more obvious in this image.

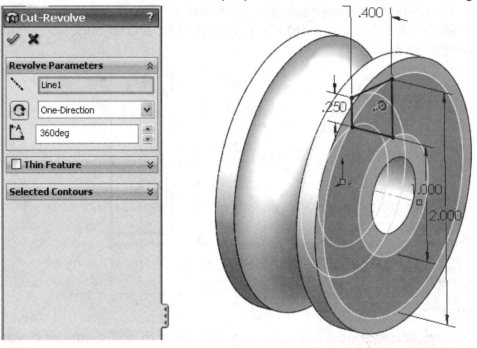

134. - Since the cut has to be on both sides of the part, to add the second cut we'll use the "**Mirror**" command to copy the previous "Cut Revolve" about the "Front Plane" as we've done before.

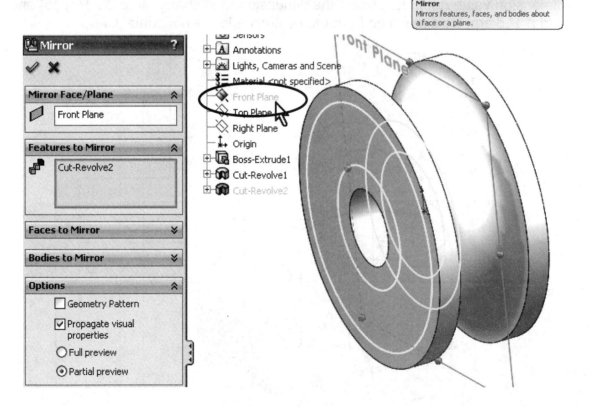

135. - To eliminate the sharp edges on the outside perimeter we'll add a 0.1" x 45° "**Chamfer**". The chamfer is an applied feature similar to the Fillet, but instead of rounding the edges, it ads a bevel to it. Select it from the drop down menu under Fillet in the Features tab or in the menu "Insert, Features, Chamfer".

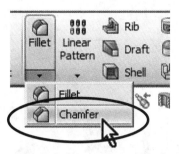

Set the options and values shown and select the two edges indicated. Click OK when finished.

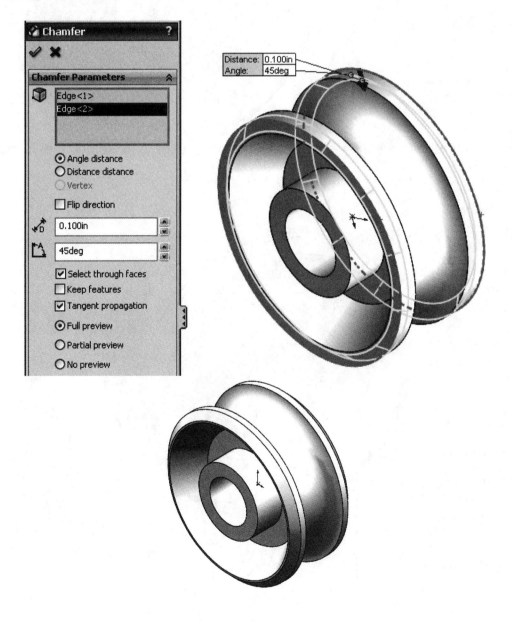

136. - Add a 0.0625" radius fillet to the four inside edges.

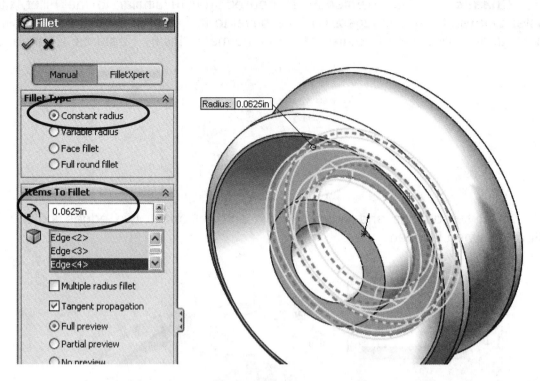

 Instead of selecting the four edges, the user can select the 2 inside faces.

137. - For the last step, we'll make the keyway. Switch to a "Front View" and make a sketch in the "Front Plane". Draw a rectangle and add a Midpoint relation between the bottom line and the origin.

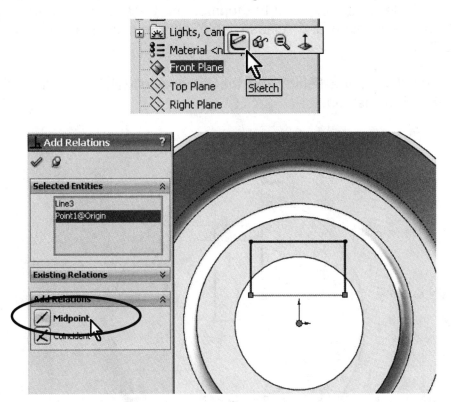

138. - By adding the Midpoint relation, the rectangle will be centered about the origin. Add a 0.188" width dimension as indicated. If the blue line (top horizontal) is below the center hole's circle, drag it until it's above the circle as shown.

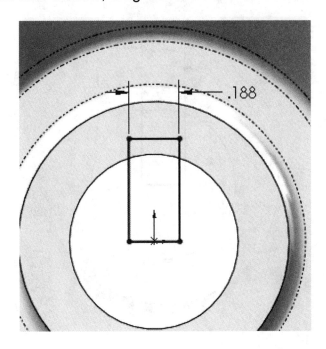

139. - Now add a dimension from the top of the circular edge to the horizontal line. Before adding the dimension, <u>press and hold</u> the "Shift" key, by doing this we'll be able to add the dimension from the top edge of the circle to the top line of the rectangle. Select the circle at the topmost part of the circular edge and the line, click for the location and release the "Shift" key; change the dimension's value to 0.094".

 If we don't hold the "Shift" key the dimension will be referenced to the circle's center instead.

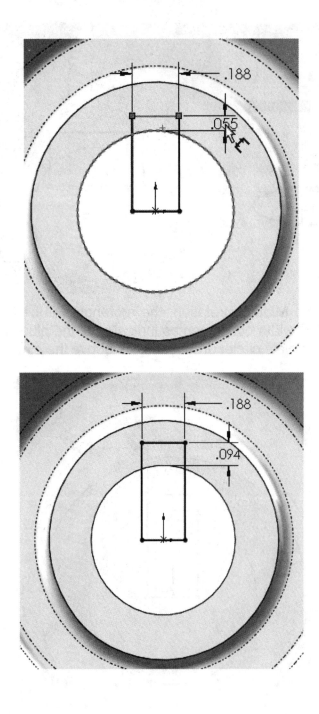

140. - Since the sketch is located in the "Front Plane", making a cut "Through All" will go only from the middle of the part to one side. What we need to do is to activate the "Direction 2" checkbox, and set the end condition to "Through All" for both "Direction 1" and "Direction 2" to finish the part.

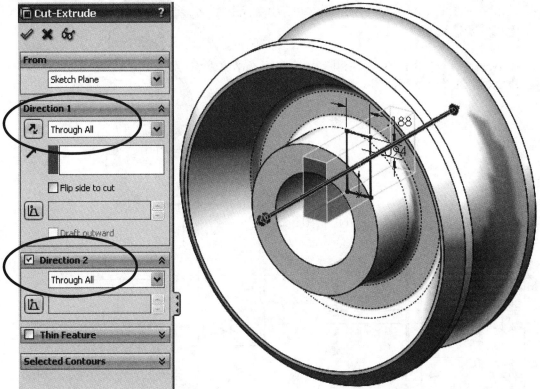

 We can use any End Condition needed for Extrude or Cut features in either direction as required.

141. - Change the material to "AISI 1020" steel. Save the part as "Worm Gear" and close the file.

Exercises: Build the following parts using the knowledge acquired in this lesson. Try to use the most efficient method to complete the model.

Exercise 11

Units: inch

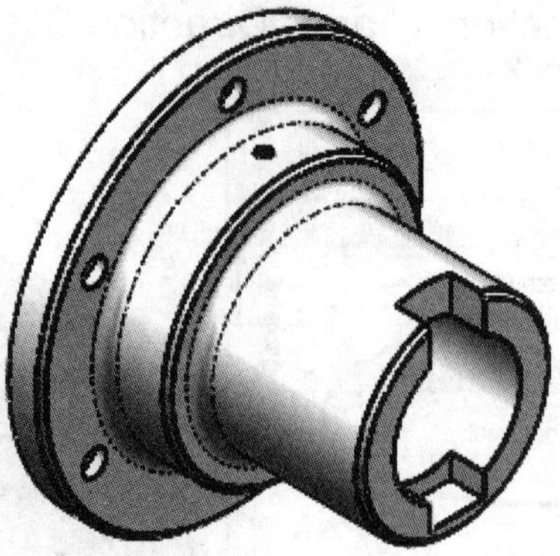

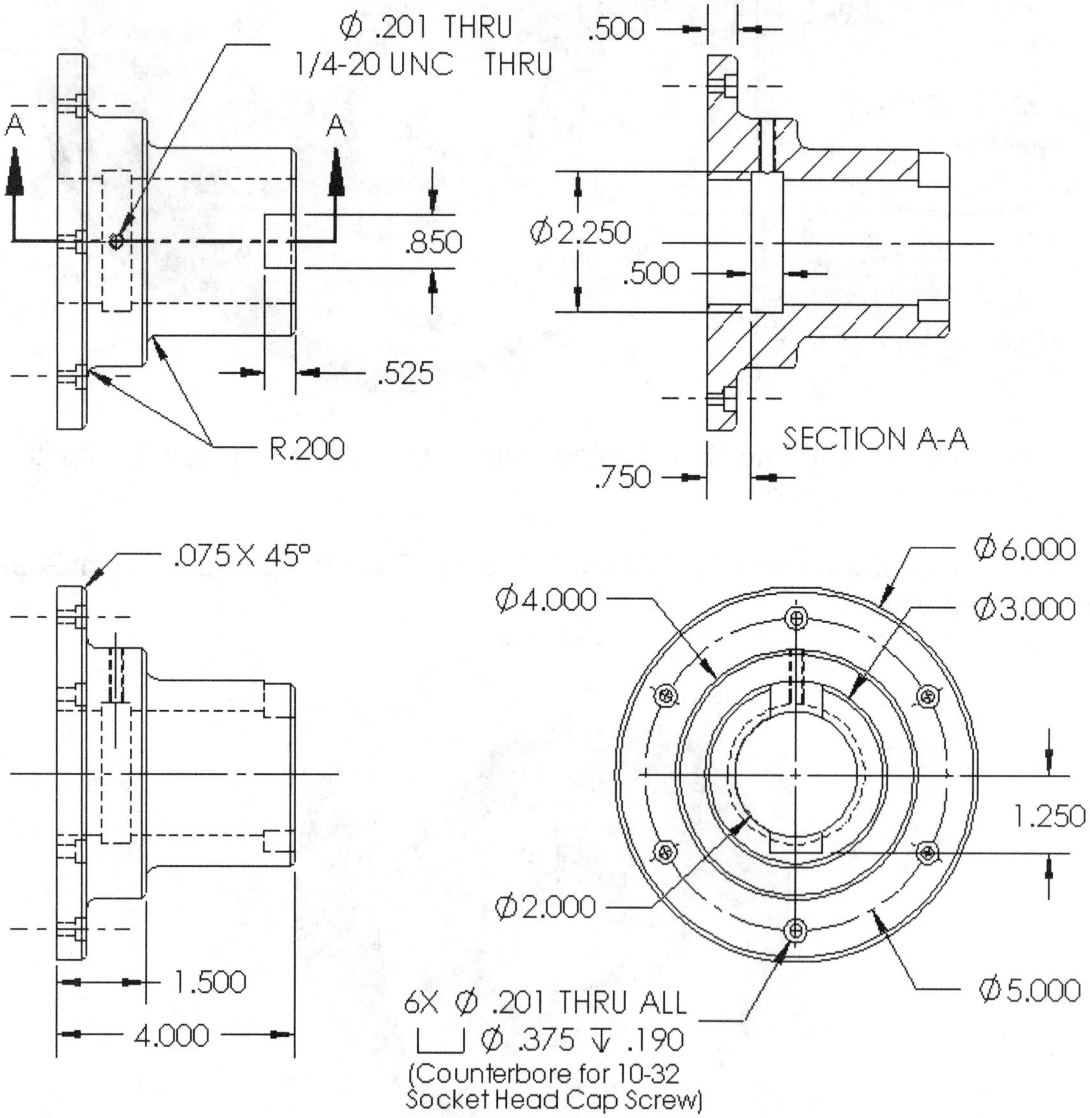

Ø .201 THRU
1/4-20 UNC   THRU

.500

.850

.525

R.200

Ø2.250

.500

SECTION A-A

.750

.075 X 45°

Ø4.000

Ø6.000

Ø3.000

1.250

Ø2.000

1.500

4.000

6X Ø .201 THRU ALL
⊔⊔ Ø .375 ▽ .190
(Counterbore for 10-32
Socket Head Cap Screw)

Ø5.000

The Worm Gear Shaft

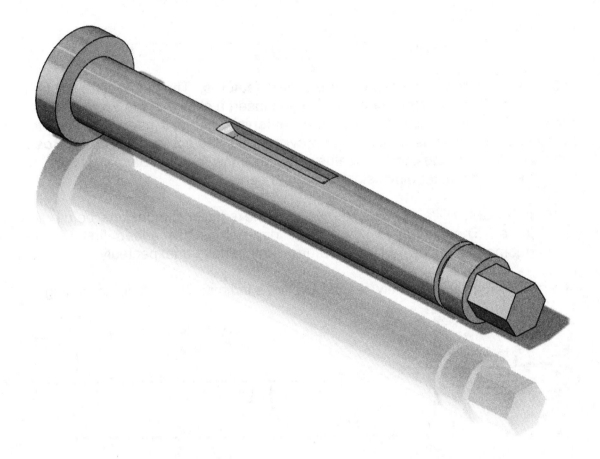

For the Worm Gear Shaft we will review the Revolved Feature previously learned, the sketch Polygon tool and a Mid Plane cut.

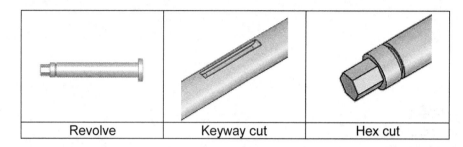

| Revolve | Keyway cut | Hex cut |

142. - In this part we only need to make three features. The first feature will be a revolved extrusion. Make a new part file and insert a sketch in the Front plane as shown. It's very important to add the centerline, as we'll need it to add the diameter dimensions as we did in the previous part. Select the "**Revolved Boss**" command to complete the first feature. Rename the feature "Base". Notice the three doubled diameter dimensions using the centerline.

 In a sketch like this is better to add the smaller dimensions before the larger. The reason is if the large dimensions are added first, as the geometry updates the small features may behave unexpectedly.

Add a "Collinear" relation between the two lines indicated before dimensioning.

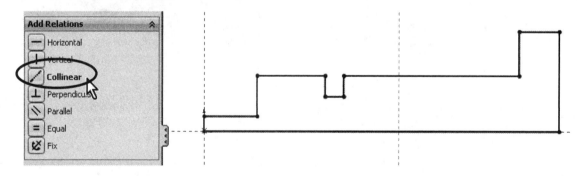

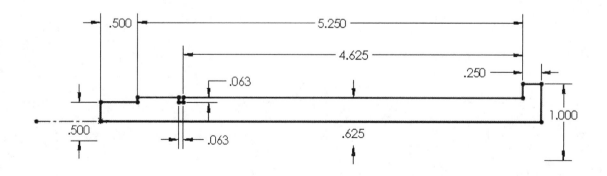

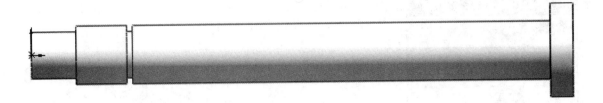

143. - The second feature is the keyway. Select the "Front Plane" from the Feature Manager and click in the "**Sketch**" icon from the pop up toolbar. In this sketch it's OK to leave the top line of the sketch under defined. What we'll do is a cut using the "**Mid Plane**" end condition; the top line beyond the top of the part allows us to "cut air", this way we are sure we will cut the part. This is a common practice and is fine as long as we make sure the sketch is big enough to accommodate possible future changes.

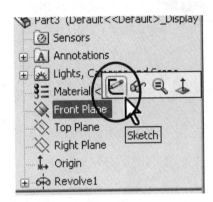

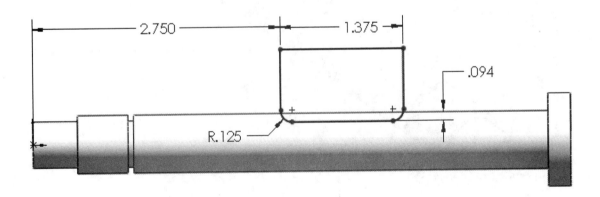

144. - Select the "Cut Extrude" command from the "Features" tab and use the "**Mid Plane**" end condition with a dimension of 0.1875".

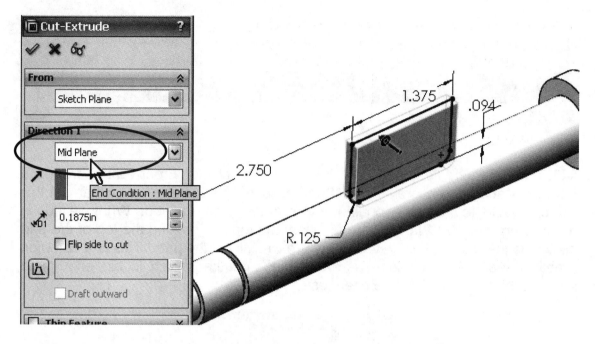

Finished keyway cut.

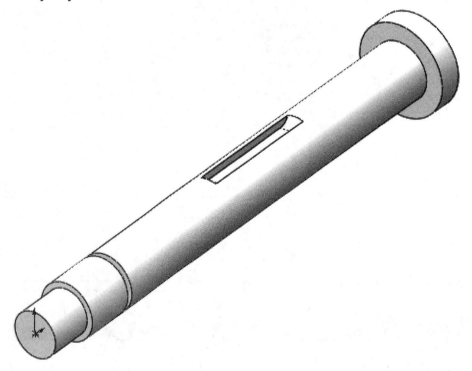

145. - For the last feature we'll make a hexagonal cut just as we did in the "Offset Shaft". Follow the same steps we did with the "Offset Shaft", with the exception that we are making the hexagonal cut in the left most face of the part. Switch to a "Left View", select the leftmost face and create a sketch. Using the "**Polygon**" tool from the "Sketch" tab draw a Circumscribed hexagon and make the construction circle equal to the shaft. Finally make one line horizontal to fully define the sketch.

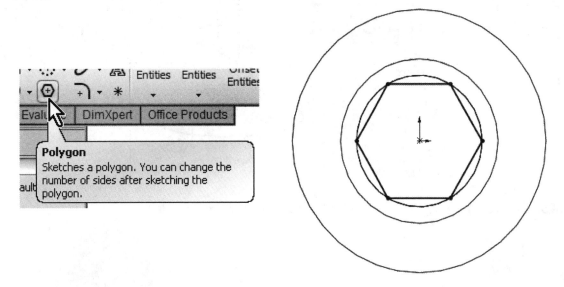

146. - Make a "Cut Extrude" using the "Up to Surface" end condition as indicated and activate the option "Flip side to cut" to cut *outside* the hexagon.

 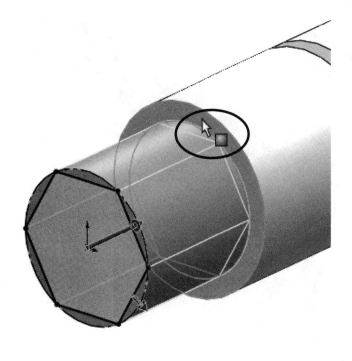

147. - Change the material to "Chrome Stainless Steel" as the final step.

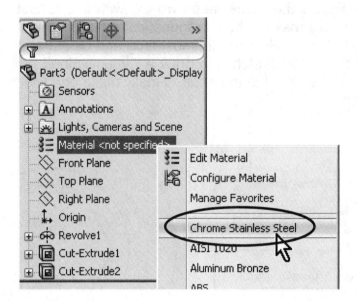

148. - Save the part as "Worm Gear Shaft" and close the file.

Special Features: Sweep and Loft

There are times when we need to design components that cannot be easily defined by prismatic shapes. For those features that have 'curvy' shapes we can use **Sweep** and **Loft** features, which let us create almost any shape we can think of. These are the features that allow us to design consumer products that, by definition, have to be attractive and are frequently made of plastic. These products include things like your remote control, a computer mouse, bottles, phones, etc. Many times these products' appearances can be their success or failure in the market. They have to look nice, 'feel' right, and above all work well. Sweeps and Lofts are also prevalent in the automotive and aerospace industry where cosmetics, aerodynamics and ergonomics are a very important part of the design.

Sweeps and Lofts have many different options that allow us to create from simple to extremely complex shapes. In light of the vast number of variations and possibilities for these features, we'll keep these examples as simple as possible without sacrificing functionality, and above all give the reader a good idea as to what can be achieved.

Sweeps and Lofts are usually referred to as advanced features, since they usually require more work to complete, and a better understanding of the basic concepts of solid modeling. Having said that, these exercises will assume that any command that we have done more than a couple of times up to this point, like creating a sketch are already understood and we'll simply direct the reader to create it providing the necessary details. This way we'll be able to focus more in the specifics of the new features that we are learning. These are some examples of consumer product designs made using advanced modeling techniques.

Notes:

Sweep: Cup and Springs

For this exercise we are going to make a simple cup. We will learn a new option when creating features called "Thin Feature", Sweep feature and a new fillet option to create a Full Round fillet as well as a review of creation of auxiliary Planes. The sequence of features to complete the cup is:

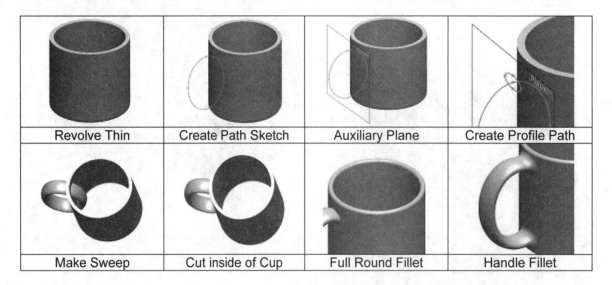

| Revolve Thin | Create Path Sketch | Auxiliary Plane | Create Profile Path |
| Make Sweep | Cut inside of Cup | Full Round Fillet | Handle Fillet |

149. - For the first feature we will create a "Revolved Feature" using the option "Thin Feature". The thin feature makes a feature with a specified thickness based on the sketch that was drawn. Select the "Front" plane and create the following sketch. Notice the sketch is only two lines, an arc and a centerline.

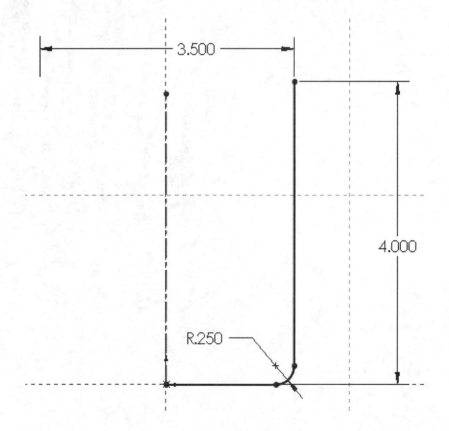

150. - After selecting the "Revolve" command we get a warning telling us about the sketch being open. Since we want a thin revolved feature, select "No".

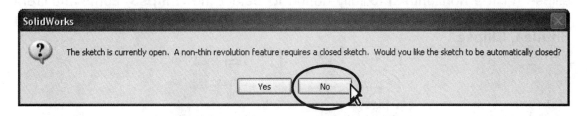

In the "Revolve" options, **"Thin Feature"** is automatically activated. For the cup we want the dimensions we added to be external, so we select the "Reverse Direction" option. Notice the preview shows the change adding material *inside*.

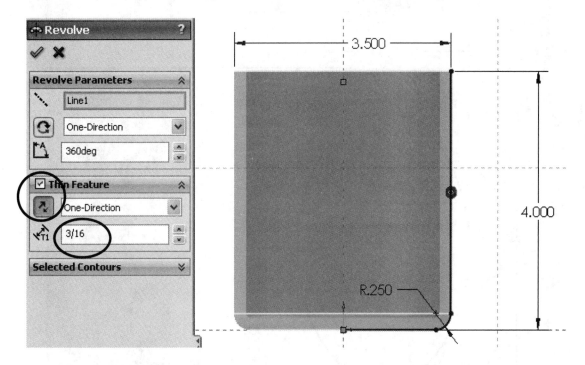

 In the value box we typed 3/16; we can add a fraction and SolidWorks changes it to the corresponding decimal value when we click OK; we can also type simple mathematical expressions with addition, subtraction, multiplication and division in any box where we can type a value.

151. - Select the "Front" plane and create the following sketch using an ellipse. Be sure to add a "Vertical" relation between the top and bottom points of the ellipse (or the horizontals) to fully define it. Select the "**Ellipse**" command from the Sketch tab in the Command Manager or from the menu "**Tools, Sketch Entities, Ellipse**".

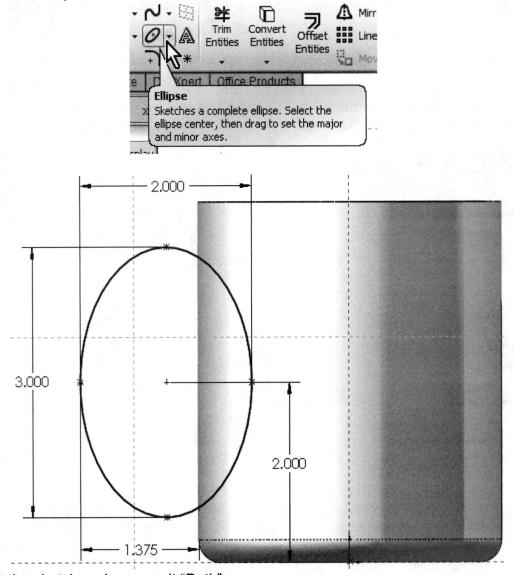

Exit the sketch and rename it "Path".

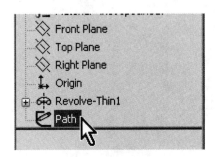

152. - Create an Auxiliary plane parallel to the "Right" plane using the center of the ellipse as a second reference as shown.

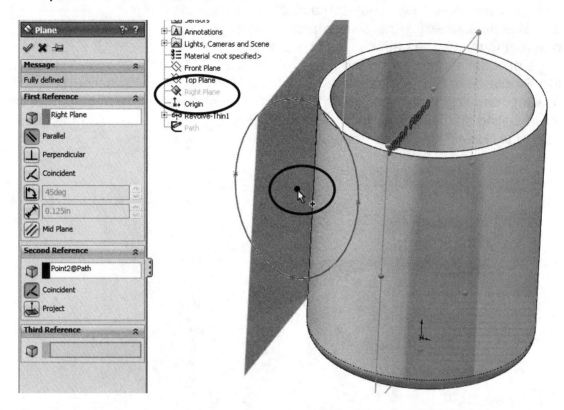

153. - Select the plane just created and draw the next sketch. Start the center of the ellipse at the top point of the ellipse in the previous sketch. Remember to add a horizontal relation between the points of the major axis. Exit the sketch and rename it "Profile".

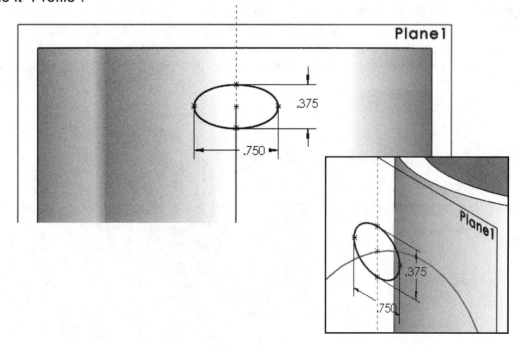

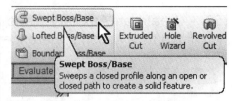

154. - Select the "**Sweep**" icon from the "Features" tab in the Command Manager. The sweep is a feature that needs to have at least two sketches, one for the sweep profile, and one for the path. The path can also be a model edge or a defined curve. In the "Sweep" properties select the "Profile" sketch in the Profile selection box, and the "Path" sketch in the Path selection box. The "Sweep" can optionally have guide curves and other parameters to better control the resulting shape. Notice the preview and click OK when done. Hide the planes using the "Hide/Show Items" toolbar.

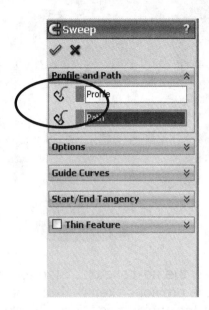

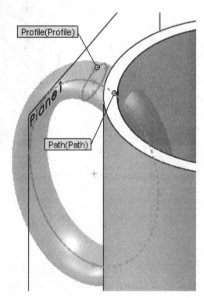

155. - Notice the sweep is also inside the cup. We will make a cut to fix this. Create a sketch in the flat face at the top of the cup. Select the inside edge at the top and use "Convert Entities" from the "Sketch" tab to convert the edge to sketch geometry. Select the "Extruded Cut" icon and use the option "Up to next" from the end condition drop down selection.

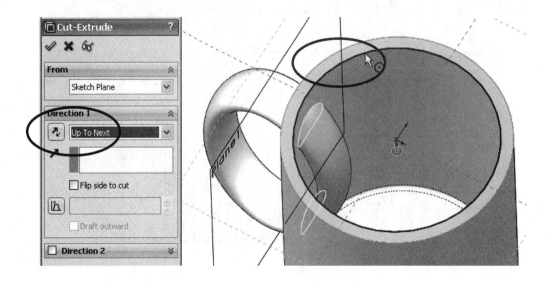

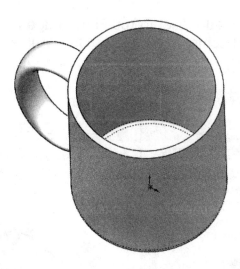

156. - Now we need to round the top of the cup. To do this we'll use the "Fillet" command using the option "**Full Round Fillet**" from the "Fillet Type" options. There are three faces that we need to select, the middle face will be replaced by the fillet. The first selection box is active; select the outside face of the cup. Click inside the second selection box (Center Face) and select the top face of the cup. Click inside the third box and select the inside of the cup. Click OK to finish.

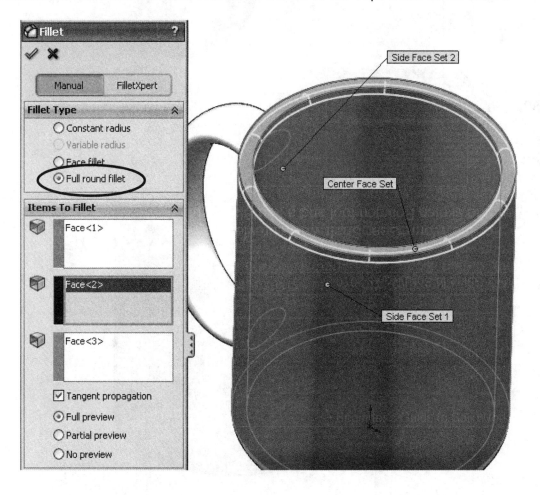

157. - To finish the cup add a fillet to round the edges where the handle meets the cup. Select the fillet command with the "Constant Radius" option, select the handle surface and change the radius to 0.25". Click OK to finish.

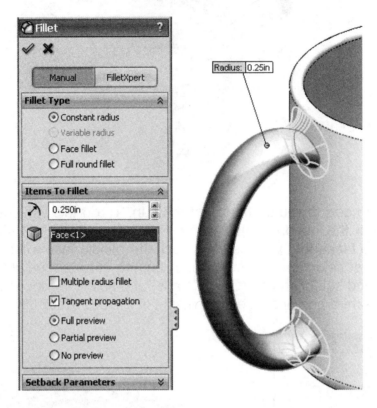

158. - Save the part as "Cup" and close.

In this exercise we are going to show how to make a simple and a variable pitch spring. In order to make these springs we'll learn how to make a simple as well as a variable pitch helix to be used for the sweep's path. The sequence of features to complete the springs is:

Simple Spring

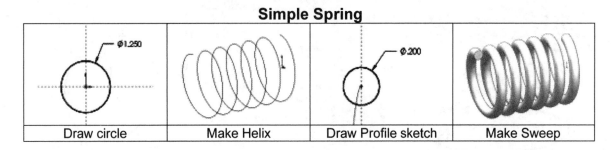

| Draw circle | Make Helix | Draw Profile sketch | Make Sweep |

Variable Pitch Spring

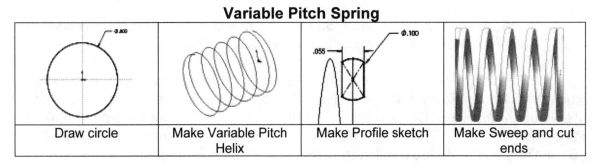

| Draw circle | Make Variable Pitch Helix | Make Profile sketch | Make Sweep and cut ends |

Simple Spring

159. - In order to make any helix we need to make a sketch with a circle first, that is going to be the helix's diameter. Select the Front Plane and make a sketch using the following dimensions and Exit the sketch.

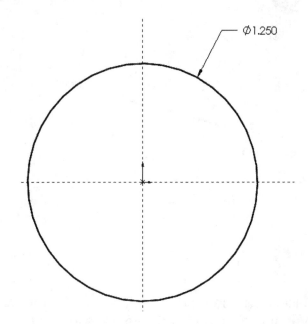

160. - In the Features tab select "Curves" then "**Helix and Spiral**" from the drop-down menu. When asked to select a plane or a sketch, select the sketch we just drew.

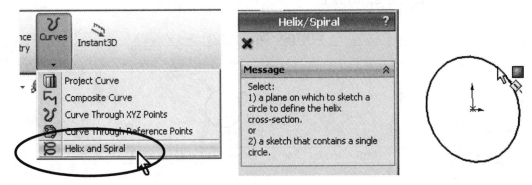

 If we select the "Helix and Spiral" command before exiting the sketch, SolidWorks will use that sketch automatically for the helix.

161. - The helix can be defined by the combination of any two parameters including Pitch, Revolutions and Height. For this example select "Pitch and Revolution" from the "Defined By:" drop down menu, and make the pitch 0.325in and Revolutions 6. The "Start Angle" value will define where the helix will start. By making it 90 degrees it will start coincident with the Front plane.

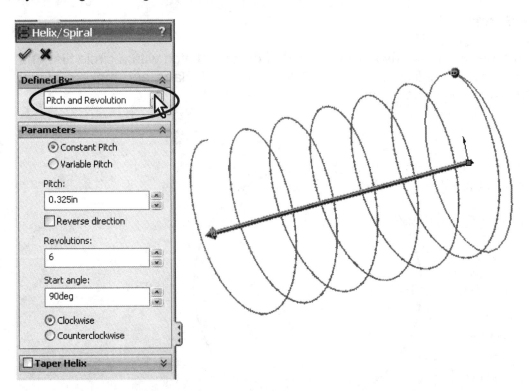

 Note that the Helix can be made Clockwise or Counterclockwise, Tapered, Constant or Variable pitch and can also be reversed (going right or left).

162. - Once the Helix is finished, we need to make the Profile sketch for the sweep. Switch to a Right View, and add a sketch in the Right Plane as shown close to the Helix.

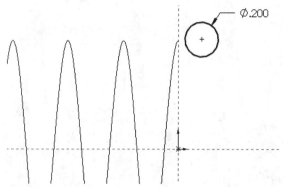

Add a "**Pierce**" geometric relation between the center of the circle and the helix, this way the path will start with the helix. This relation will make the sketch fully defined. Exit the Sketch.

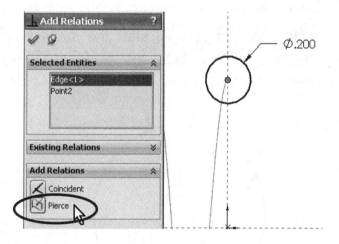

163. - Select the Sweep command and make the sweep using the last sketch as a profile and the Helix as a Path. Notice the Preview.

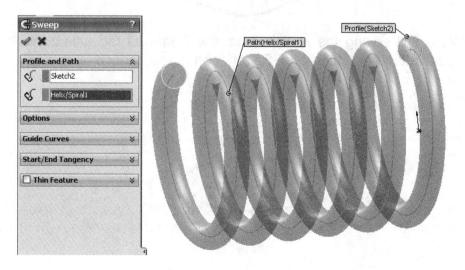

Finished Spring.

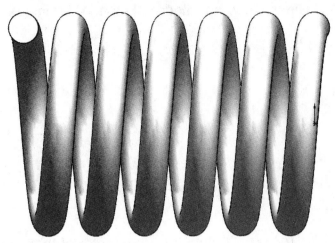

Variable Pitch Spring

164. - For the variable pitch spring we'll start the same way and make the following sketch in the Front Plane. This will be the spring's outside diameter.

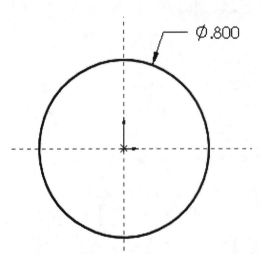

165. - While still editing the sketch, from the Features tab select the "Curves" icon and select the "Helix and Spiral" command; notice it is the only one available.

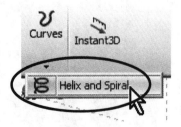

166. - From the Helix/Spiral command properties, select "Pitch and Revolution" from the "Defined by:" selection box, and "Variable Pitch" in the Parameters box. After selecting it we are presented with a table to fill in the values of the Helix. Fill in the following values in the table. Notice the preview of the helix. Click OK when finished.

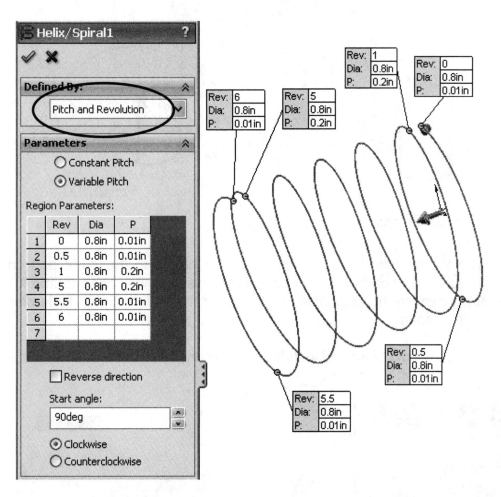

167. - After the helix is complete, add a new sketch in the Right plane, draw the circle first, add a center rectangle after, then trim and dimension as needed.

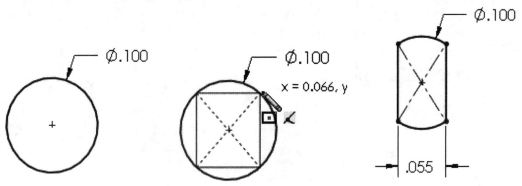

168. - Add a centerline from the center to the top line (make sure it is coincident). Select the endpoint of the centerline and the helix to add a **"Pierce"** geometric relation to make the sketch fully defined.   Exit the sketch and optionally rename it "Profile".

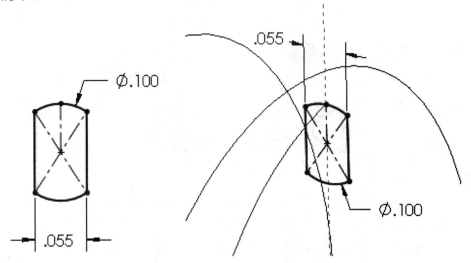

 The Pierce relation allows us to fix the profile to the path, and can be added to any part of the sketch. In this example we chose to add it to the top, because the original sketch used for the helix is the outside diameter of the spring.

169. - Just as we did before, select the Sweep command and add the path and profile. Click OK when finished.

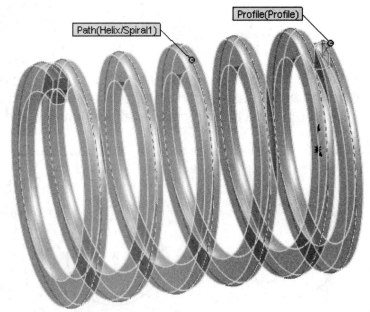

170. - As a finishing touch we'll make a cut to the sides to get the ends flat. Change to a Right view and add a sketch in the Right Plane. Add a rectangle making the start Coincident to the mid-point of the edge indicated, and to the right just past the origin.

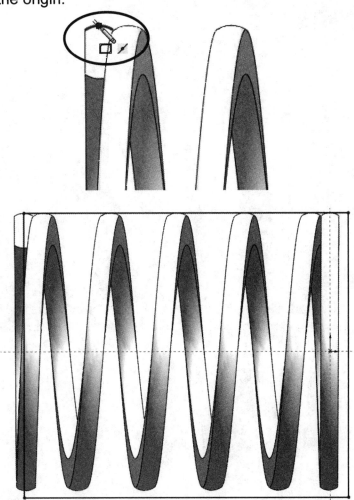

Add a Midpoint geometric relation between the right vertical line and the origin to complete the sketch.

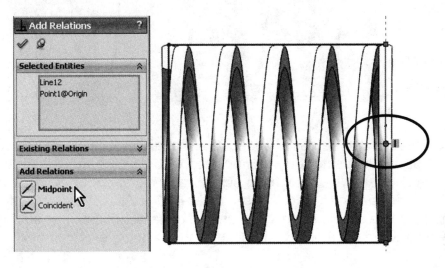

171. - Select the "Extruded Cut" command. Use the "Flip side to cut" option to cut *outside* the rectangle with the "Through All" end condition in both directions to finish.

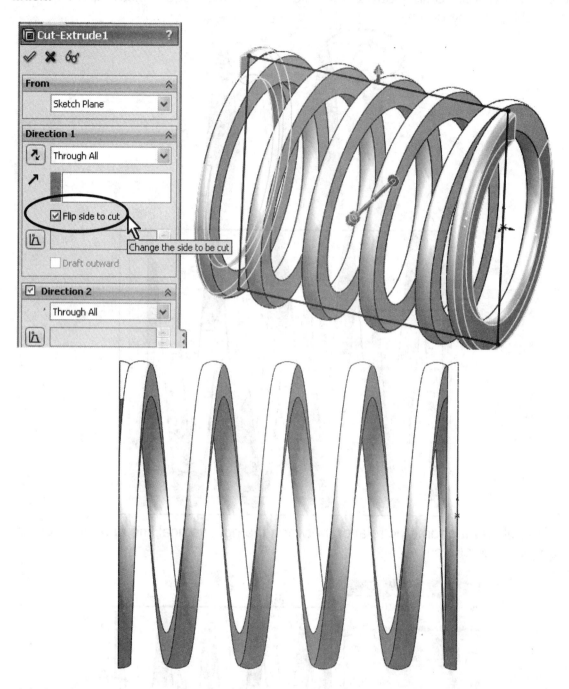

Threads can be easily added to a model using this approach, or a spiral cut made using the menu "Insert, Cut, Sweep" which will remove material. When modeling screws and fasteners in general, it is usually unnecessary to add a helical thread, as it uses a lot of computing resources, and a simple representation of it usually suffices. It is strongly advised to only do it when the model requires it, as in the bottle example next.

Exercise: Build the following bottle using the knowledge acquired in this lesson. Download the instructions from our website in the download section.

www.mechanicad.com

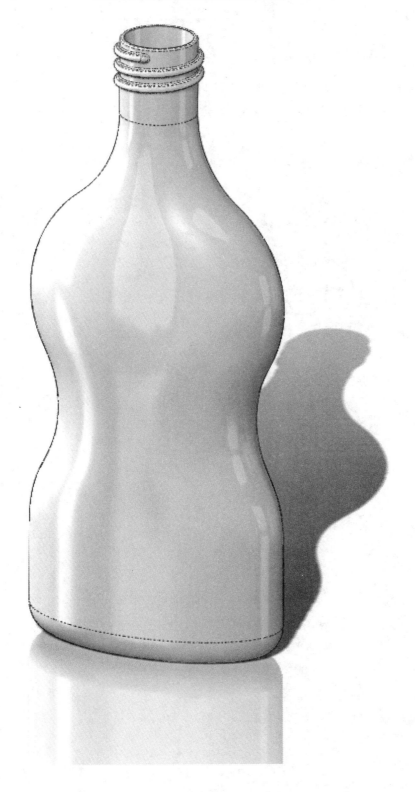

Notes:

Loft: Bottle

The "**Loft**" feature requires at least two different sketches and/or faces and optionally guide curves to more accurately define the final shape. The **Loft** helps us design more complex shapes. In this exercise we will make a bottle using a loft with four different sketches.

172. - Make a new part document and create three auxiliary planes using the "Plane" icon from the Reference Geometry drop-down menu in the "Features" tab. Select the "Top" plane as reference, change the number of planes to 3 and space them 2.5" as shown. Click OK when done.

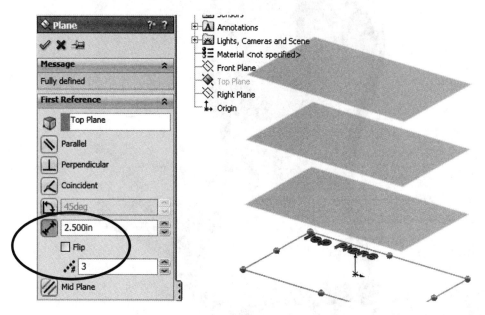

173. - For clarity, select the "Top" plane in the Feature Manager and show it using the "Hide/Show" command from the pop up toolbar.

174. - Switch to a "Top" view, select the "Top" plane and draw the following sketch using the "Center Rectangle" and "Sketch Fillet" tool. Exit the sketch when finished.

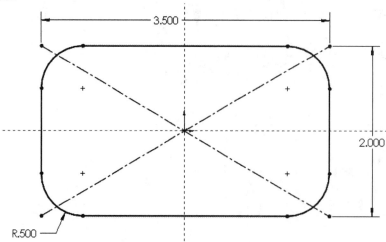

175. - Still in the "Top" view, select "Plane1" from the Feature Manager and create the second sketch, adding automatic relations to the origin and a construction line of the first sketch. Add a 2.5" dimension and the 0.5" "Sketch Fillet" to fully define the sketch. Exit the sketch when finished.

 Turn off the display of Planes for clarity with "Hide/Show Items".

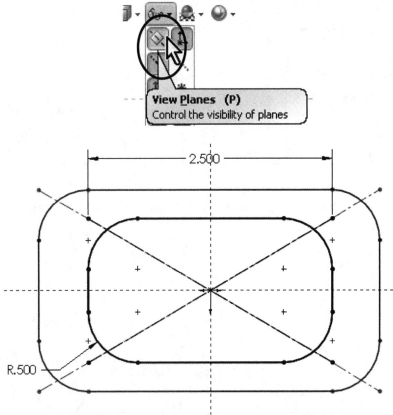

176. - For the third profile, select "Plane2" from the Feature Manager and create a new sketch. This sketch will be exactly the same as the first one. To help us save time and to keep the design intent, select the "**Convert Entities**" tool. In the "Convert Entities" selection box, select the entire Sketch1 from the Fly-out Feature Manager. Exit the sketch when finished.

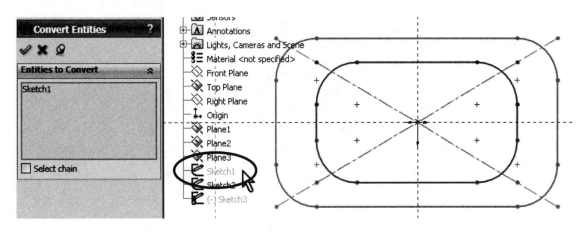

177. - For the last profile select "Plane3" from the Feature Manager and create a new sketch. Draw a circle and add a geometric relation to make it "Tangent" to the previous sketch. This will fully define the sketch. Exit the sketch to finish.

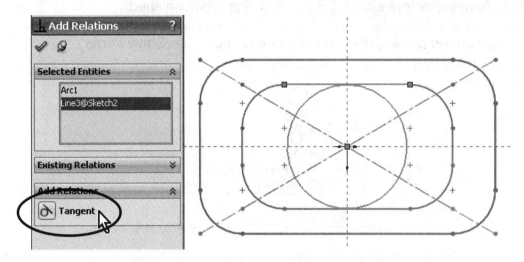

178. - The finished sketches will look like this with the planes visible:

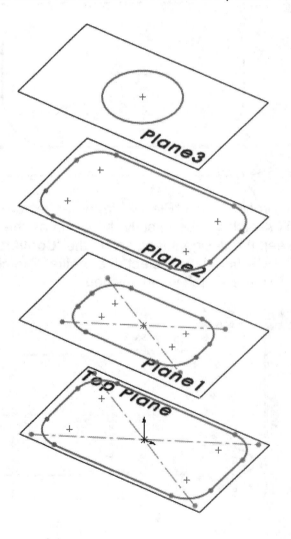

179. - Now we are ready, select the "**Loft**" icon from the "Features" tab. We will select the profiles starting with the first one we made at the bottom and finishing at the top. It is important to select the sketches thinking that where we select the profile will affect the result. Click in the graphics area near the indicated 'dots', this line indicates the segment of one profile that will be connected to the next profile. If we randomly select other points in the profile, the loft could twist and produce undesirable results. Optionally guide curves can be added to improve control of the loft.

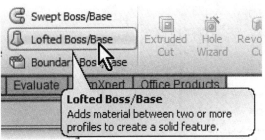

From the "Start/End Conditions" select "Normal to Profile" for both the "Start" and "End" constraint. Notice the difference in the preview after selecting the start and end constraints. Click on OK when done.

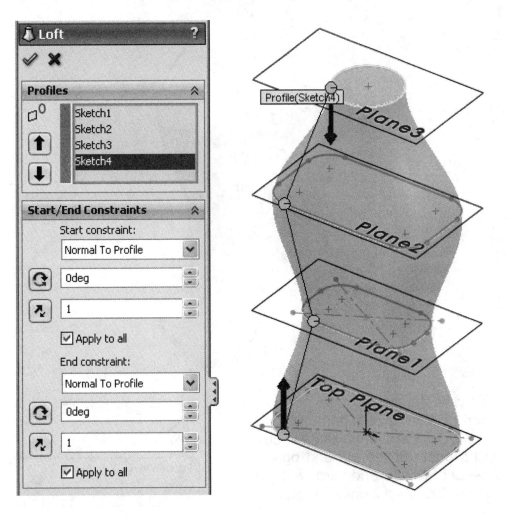

180. - Add a 0.25" radius fillet in the bottom edge of the part. Finish the part using the "Shell" tool; select the top face to remove it and make the wall thickness 0.125".

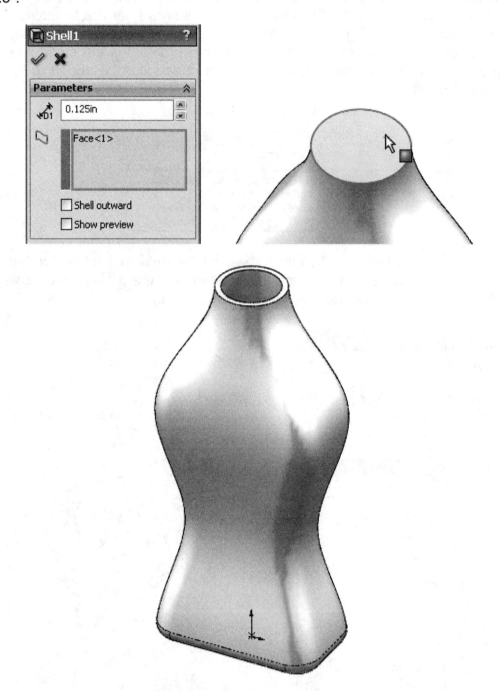

181. - To view the inside of the part, select the "**Section View**" icon from the "View" toolbar. We can change the plane to cut the model, the depth of the cut and optionally add a second section and third sections. If we click on OK when done, the model will

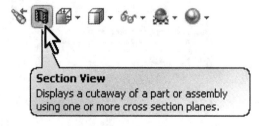

Section View
Displays a cutaway of a part or assembly using one or more cross section planes.

be displayed as cut, but this is only for display purposes, the part is not actually cut. To turn off the "Section View" select its icon again. This section view can be used along with the "Measure" tool to inspect the part.

 We can also change the depth of the cut by dragging the arrow in the center of the plane.

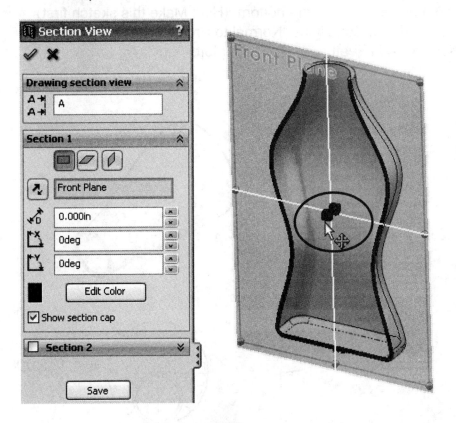

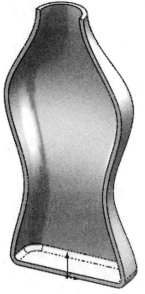

Save the part and close the file.

Exercises: Build the following parts using the knowledge acquired in this lesson. Try to use the most efficient method to complete the model.

Eccentric Coupler

Notes:

- Both circles are centered horizontally (Right view).
- Add a guide sketch at the bottom. (Hint: Make this sketch first)
- Start and End Conditions "Normal to Profile"
- Make as Thin Feature or Shell after loft.

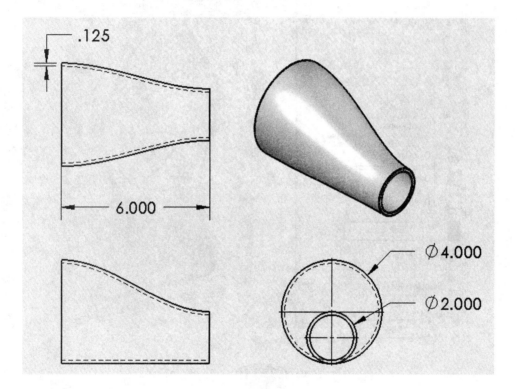

Bent Coupler

Build the following part using a Loft feature and a shell. The part is 0.15" thick.

Notes:
- Add a guide sketch along the right side of the part.
- Start and End Conditions "Normal to Profile".
- Make as Thin Feature or Shell after loft.

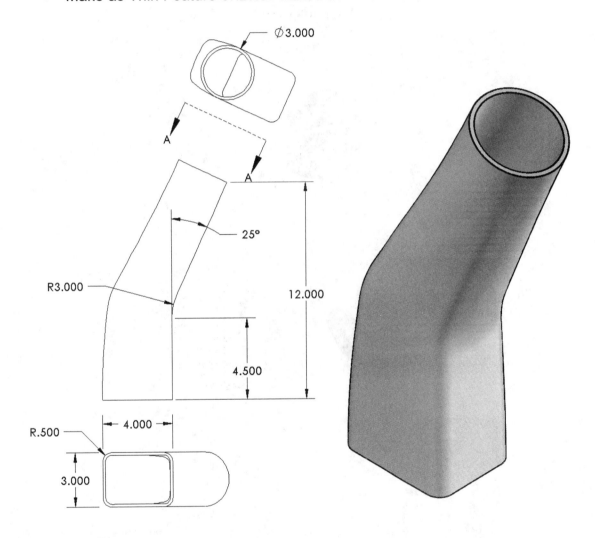

Challenge Exercises:

In order to save space (and trees), visit our web site to download the complete details for the gas grill project as well as the finished parts for this book, higher resolution images of the exercises and some extra topics not covered in the book. Build the parts and put them all together after the Assembly lesson.

www.mechanicad.com

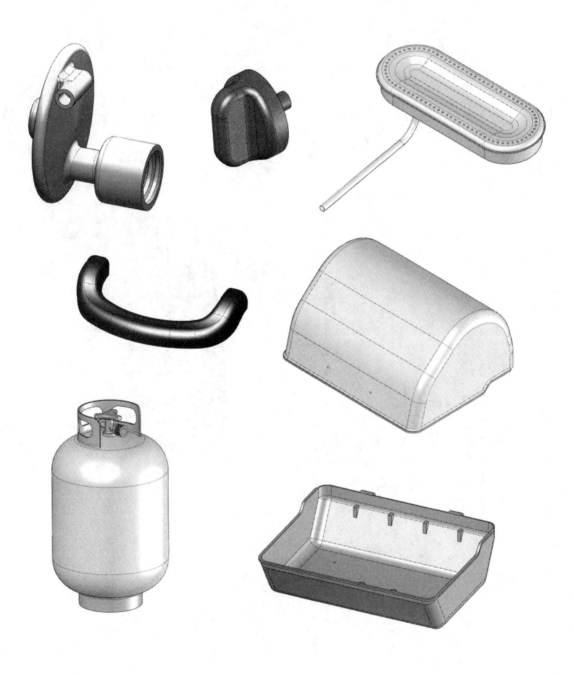

Detail Drawing

Now that we have completed modeling the components, it's time to make the 2D detail drawings for manufacturing. In SolidWorks, first we have to make the 3D models and from them derive the 2D drawings. By deriving the drawing from the solid model, the 2D drawing is linked and associated to the part. This means that if the part is changed, the drawing will be updated and vice versa. Drawing files in SolidWorks have the extension *.slddrw and each drawing can contain multiple sheets, each corresponding to a different printed page.

SolidWorks offers a very simple to use environment where we can easily create 2D drawings of parts and assemblies. In this section we'll cover Part drawings only. Assembly drawings will be covered after the assembly section. We will add different views, annotations, dimensions, sections and details necessary for manufacturing to fabricate the component without missing any detail. We will also introduce a new concept called Configurations, which allow us to show different versions of the same part, for example, a version for the part as it comes out of the foundry and a version for the machine shop with all the details for machining the finished part.

We can create our 2D drawings using any of the many dimensioning standards available in the industry. In this book we will use the ANSI standard. Once in a drawing, the dimensioning standard can be easily changed to a different one by going to the menu "**Tools, Options**" and selecting the "Document Properties" tab. In the Drafting Standard section we can simply select the desired detailing standard. It is important to note that after changing a standard, SolidWorks will change the dimension styles, arrow head type, etc. accordingly. For more information on units see the Appendix.

The detailing environment of SolidWorks is a true What-You-See-Is-What-You-Get interface. When we make a new drawing, we are asked what size of drawing sheet we want to use unless we are using a template with a predefined sheet size. This size corresponds to the printed sheet size. Do not be too concerned about selecting the right sheet size, as we can easily change to a larger or smaller sheet if we find that our drawing will not fit the current sheet.

Notes:

Drawing One: The Housing

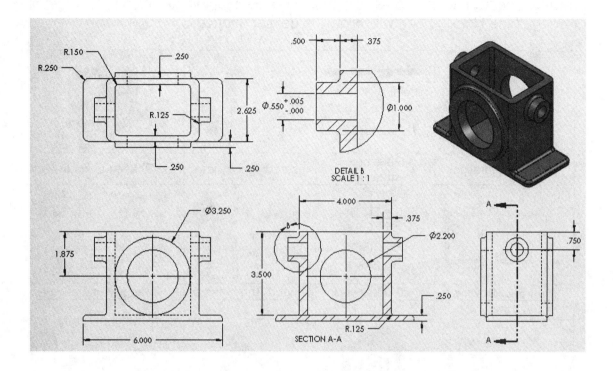

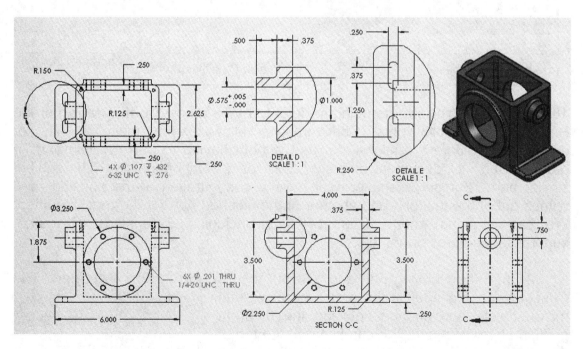

In this lesson we'll learn how to make configurations of a part, how to make a multi sheet drawing with the different configurations, add views including sections and details, change the view display style, move the views within the drawing sheet, import model dimensions and manipulate them. The detail drawing of the Housing part will follow the next sequence.

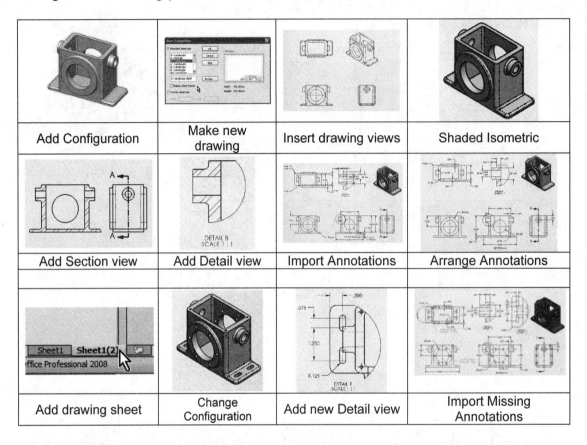

Add Configuration	Make new drawing	Insert drawing views	Shaded Isometric
Add Section view	Add Detail view	Import Annotations	Arrange Annotations
Add drawing sheet	Change Configuration	Add new Detail view	Import Missing Annotations

182. - The first thing we are going to do is to make a new **Configuration** of the Housing part. SolidWorks' Configurations allow us to make slightly or considerably different versions of a part without having to make a new part, in order to show or document different states of a component or a different but similar part. For our example, we'll have a configuration of the housing as it comes out of the foundry, and another of the finished part for the machine shop. We can configure many different things, including dimensions, tolerance, suppressed features or not, etc.

Open the Housing part and activate the "**Configuration Manager**". The Configuration manager is the area where we can make, delete and change between the different configurations of a part. We can find it at the top of the Feature Manager; it's the third tab at the top as indicated.

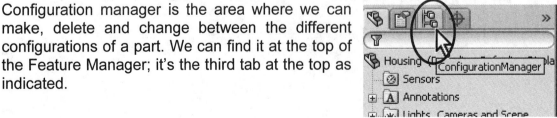

178

There is always one configuration called "Default". Once the Configuration Manager is activated, we can add a new configuration by Right-Mouse-Clicking inside the Configuration Manager and selecting the "**Add Configuration**…" option. Name the new configuration "Forge" and click OK. Adding a description is optional.

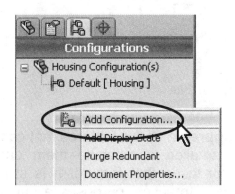

183. - Right-Mouse-Click in the "Default" configuration and select "Properties" to change its name to "Machined" and click OK.

 Configurations can also be renamed using the slow double click method.

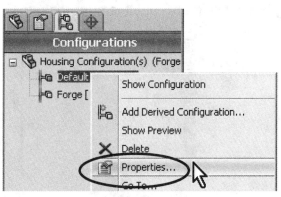

184. - The yellow colored configuration is the currently active configuration in the Configuration Manager. To switch to a different configuration, simply double click its name or Right-Mouse-Click and select "Show Configuration".

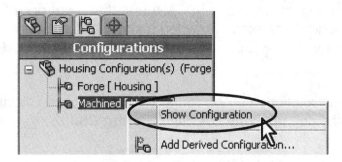

185. - Activate the Forge configuration (its name will be yellow), and switch to the Feature Manager tab. Notice that beside the name of the part at the top of the Feature Manager we can see the name of the currently active configuration.

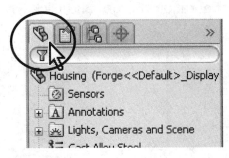

In the Forge configuration, we'll "**Suppress**" all the features that cannot be done in a forge, and make the holes smaller to machine them out later. Suppressing a feature tells SolidWorks *not* to do that feature. It is still in the Feature Manager, but it's grayed out. For all practical purposes a suppressed feature doesn't exist. Suppressing a feature affects the part's mass properties.

186. - First we will select "1/4-20 Tapped Hole1" from the Feature Manager, which cannot be made in a foundry, and from the pop up toolbar select the "**Suppress**" icon. After suppressing a feature its name will be in grayed out. Remember that **we are not deleting it**, it's still there, but it is suppressed.

Notice that the following two features, "CirPattern1" and "Mirror3" were also suppressed. This is because these features are 'children' of the "1/4-20 Tapped Hole1". Look at it this way: If there is no tapped hole, we cannot pattern it, and if there is no pattern, we cannot mirror it either. The "Parent/Child" relations are created when we add features that use or reference existing features in the model. For example, when a sketch is created in a face, the face becomes a parent to the sketch. If the face is deleted, so will be the sketch and so on.

 Parent/Child relations can be found by Right-Mouse-Clicking on a feature and from the pop up menu selecting "**Parent/Child**".

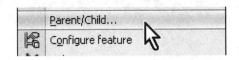

187. - Now suppress the "#6-32 Tapped Hole1" and the "Slot" features. All the children features of "Slot" are suppressed as well.

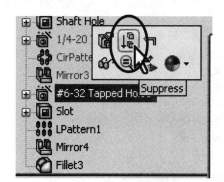
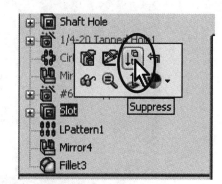

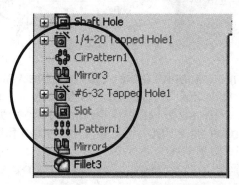

188. - The reverse process to Suppress is "**Unsuppress**". Simply select a suppressed feature, and click on the "Unsuppress" icon (the icon with the arrow pointing up).

 When Unsuppressing a feature with children in the Feature Manager, the child features may need to be Unsuppressed individually. To Unsuppress a feature and all its children, use the menu "**Edit, Unsuppress with Dependents**". If a Child feature is unsuppressed, all of the needed parent features for it to exist will be unsuppressed at the same time.

189. - The next thing we want to do is to change the value of the two circular cuts for the Forge configuration. Make sure the Forge configuration is active. If the "**Instant 3D**" command is active, select a face of the first cut in the screen. If Instant3D is not active, double click in the feature in the Feature Manager or one of its faces in the graphics area to reveal its dimensions. Right-Mouse-Click in the diameter dimension and select the "Configure Dimension" option from the pop-up menu.

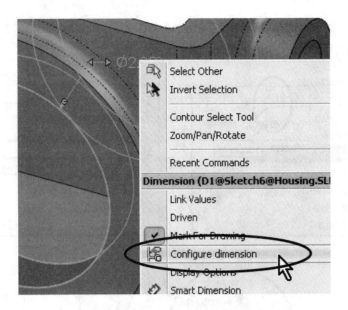

In the "Modify Configurations" table, change the value of the diameter in the "Forge" configuration to 2.200". The idea is to have a smaller hole made in the forge, and machine it to size later (Machined configuration). Leave the Machined configuration value as 2.500". Click "OK" when finished.

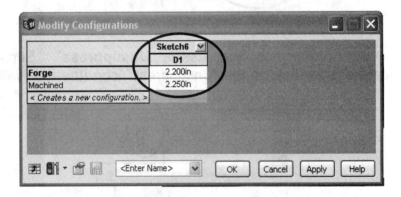

Do the same with the "Shaft Hole" and change the size for the "Forge" configuration to 0.550". Click "OK" when finished.

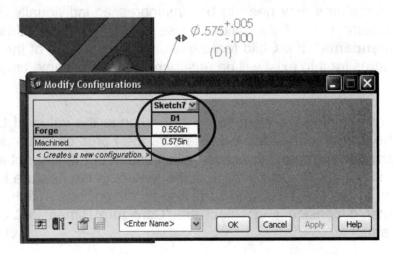

190. - An easy way to work with configurations is by viewing both the Feature Manager and the Configuration Manager at the same time, by splitting the Feature Manager pane. Move the mouse pointer to the top of the Feature Manager and look for the "split" feedback icon, then click and drag down. We can view Features in the top pane and Configurations in the bottom pane.

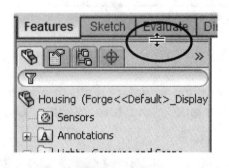

Now that we have made some changes to the "Forge" configuration, double click on each configuration name to activate it and see what each one of them looks like. Notice the missing features (suppressed) and different sizes of the holes in the "Forge" Configuration. Save these changes to the "Housing" part.

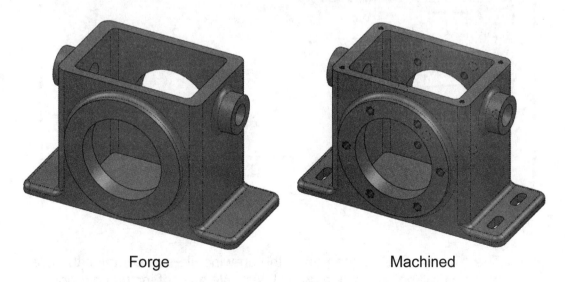

Forge   Machined

SolidWorks allows the user to add as many configurations as needed, but when we have more than 2 or 3 configurations it may be easier to manage them with a table. Design Tables is a very powerful and extensive topic. We'll cover the basics of a design table later in the book.

191. - Now, in order to make the detail drawing of the Housing part, make sure the "Forge" configuration is active, and then select the menu "**File, Make Drawing from Part**". This way we'll make the drawing using the "Forge" configuration first. We'll make the "Machined" configuration drawing after that.

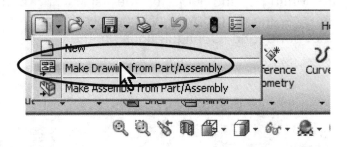

192. - We are now presented with the option to select a sheet size to use for this drawing. Select Paper Size "B-Landscape", and turn off the "**Display Sheet Format**" option.

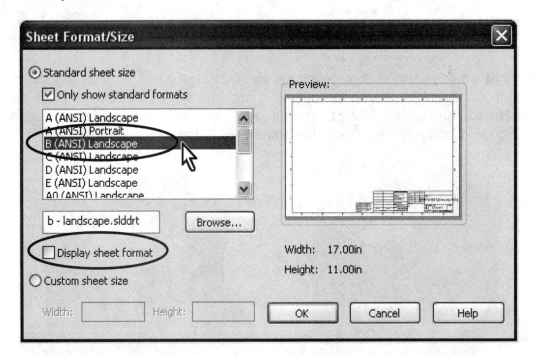

 The "sheet format" is the part of the drawing file that contains the title block. It will be covered in detail later. We are not including it now for easier visualization of this drawing.

193. - After selecting the sheet size, a new drawing is opened and SolidWorks is ready for us to choose the views that we want in the drawing. The "**View Palette**" is automatically displayed. Make sure the "Auto-start projected view" option is active, as this option will save us a little time.

194. - From the "**View Palette**" drag-and-drop the "Front" view onto the sheet.

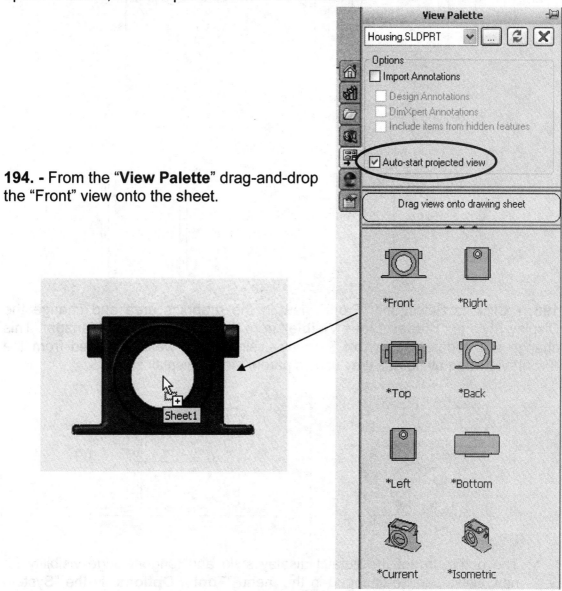

195. - Locate the "Front" view approximately in the lower left part of the sheet where we have enough room to add the other views. After locating the "Front" view on the sheet, SolidWorks automatically starts the "**Projected View**" command; simply move the mouse in the direction of the other views needed, and click to locate them on the sheet. As a guide, after adding the "Front" view, move the mouse pointer above the "Front" view and click to locate the "Top" view (you will be able to see a preview). Then click above to the right where the "Isometric" view will be and finally click to the right to locate the "Side" view. When we are done adding views, click "OK" to finish. Your drawing should now look approximately like this:

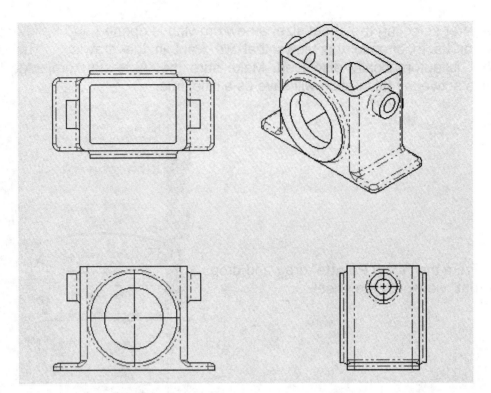

196. - Click to Select the "Front" view in the graphics area and change the Display Style to **"Hidden Lines Visible"** from the view's Property Manager. This change will update the rest of the views, since they were projected from the "Front" view and by default they use its parent view's display style.

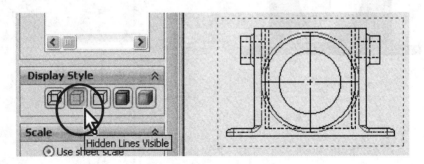

 The option to set the default display style and tangent edge visibility for new views can be changed in the menu "**Tools, Options**" in the "System Options" tab under "Display Style".

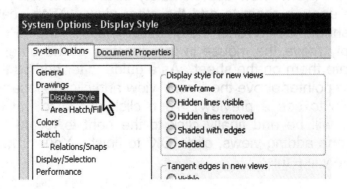

Adding standard and projected drawing views

When we make a new drawing from a part using the menu "Make Drawing from Part", SolidWorks automatically displays the "View Palette". However, if we make a new drawing using the "New Document" icon, we get a different behavior. To test it, select the "New Document" icon and create a Drawing by selecting the "Drawing" template using the same settings as before. In this case, we are presented with an empty drawing.

To activate the View Palette, click in the View Palette icon from the Task Pane on the right side of the screen to reveal it.

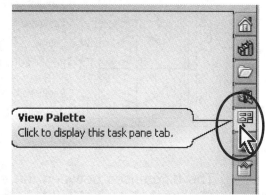

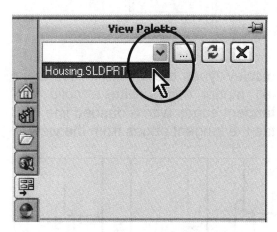

Once the "View Palette" is displayed, we can browse for a part or assembly or select one of the open documents from the drop-down list. Then it is the same behavior to drag the views onto the sheet as we did before.

197. - Now that we have the views in place, we want to show the Isometric view in "**Shaded with Edges**" mode. Click to select the "Isometric" view on the screen, and click the "**Shaded with Edges**" icon either in the "Display Style" toolbar or in the Property Manager as in the previous step. Notice the green dotted line around the view; this is indicating to us that the view is selected. Using this procedure, we can change any drawing view to any display mode.

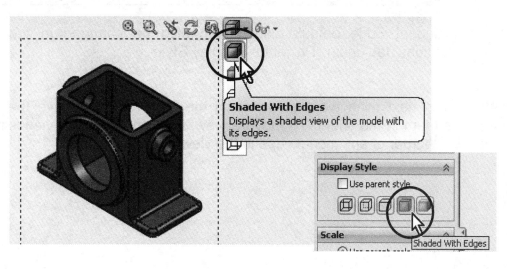

198. - In the drawing views we may or may not want to see the tangent edges. A tangent edge is where a round face and a flat face meet, such as in fillets. SolidWorks has three different ways to show them: Visible, With Font, or Removed. Right-Mouse-Click inside the "Front" view and from the Drop down menu select the option "**Tangent Edge, Tangent Edges Removed**". Repeat for the Top and Right views.

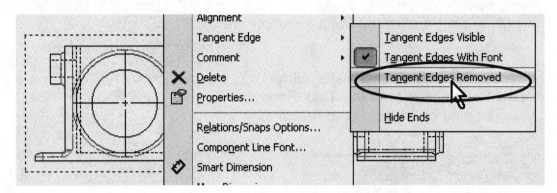

 The differences between the edge display types are visible in this table. "**Tangent Edges Visible**" shows all model edges with a solid line, "**Tangent Edges with Font**" shows tangent edges with a dashed line, and "**Tangent Edges Removed**" eliminates the tangent edges from the view.

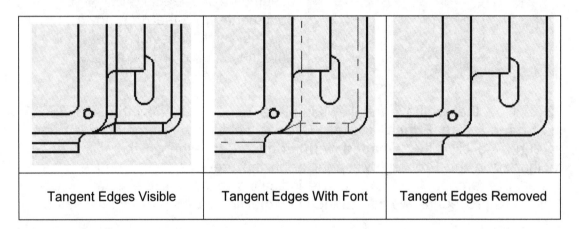

Tangent Edges Visible	Tangent Edges With Font	Tangent Edges Removed

 Tangent Edge default display can also be changed in the "System Options" tab under "Drawings, Display Style".

199. - The next thing we want to do in the drawing is to move the views to arrange them in the sheet. To move a view, click and drag the view either from the View border or any model edge in the view.

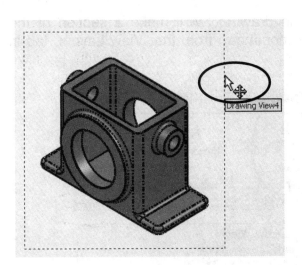

200. - Click and drag the drawing views in the drawing sheet and arrange them as shown, with "**Tangent Edges Removed**" for all views, and the Isometric with "**Shaded with Edges**" display. This layout will allow enough space to import the dimensions into the drawing.

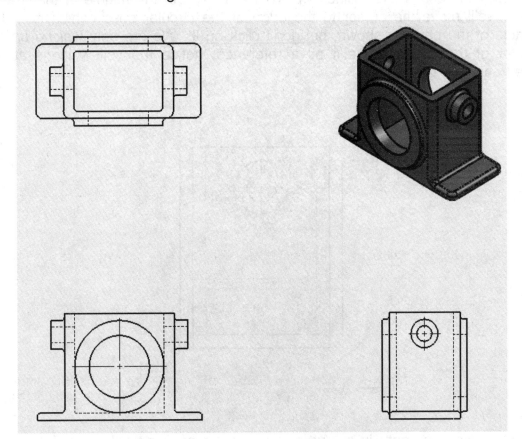

Notice the toolbars available in the Command Manager were automatically changed to match the drawing environment in which we are now. The "Features" tab was replaced by "View Layout" and "Annotation", with tools better suited for the detailing environment.

201. - To help us better understand the drawing, we'll make a section of the "Right" view. Select the "**Section View**" command from the "View Layout" tab in the Command Manager.

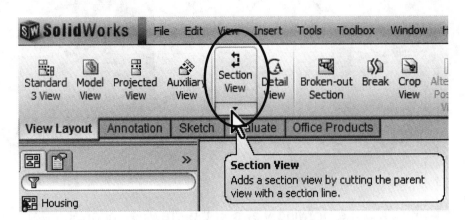

202. - SolidWorks automatically activates the "**Line**" tool and we are now ready to draw the section line. Since we want a section in the middle of the "Right" view, we'll move the cursor to the edge of the circular boss (which is in the middle of the part) as shown, but don't click on it!. We are only *waking up* the center of the circle to use it as a reference; notice the yellow icons at the quadrants and the center.

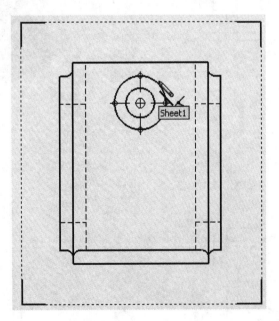

 We can zoom in or out in the Drawing environment using the mouse scroll wheel. SolidWorks will zoom in where the mouse pointer is when scrolling.

203. - Now, move the cursor above the view, and using the center of the edge as a reference, draw a line starting *above* the view and all the way across through the "Right" view as shown.

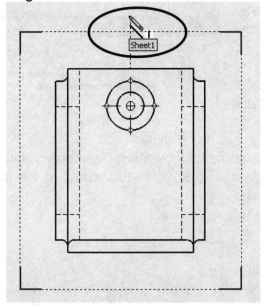 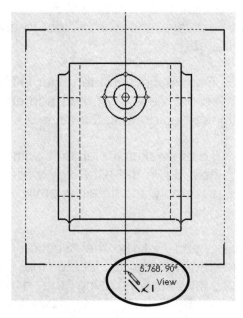

Section Line Start   Section Line End

204. - Immediately after we finish drawing the line, move the mouse and you'll see a dynamic preview showing the section of the Housing; the last thing to do is to locate the section in the sheet. Move the mouse to the left and click to locate the section between the "Front" and the "Right" views. If needed, after locating the "Section" view move the other views to arrange and space them.

 Another way to make the section view is to draw the section line first, pre-select it, and then select the "Section View" icon.

By selecting the Section View we can see it's Properties, where we can change different options such as the Section Label, reverse the section direction ("Flip Direction"), Display Style, Scale, etc. Change the Section View's display style to "**Hidden Lines Removed**" for clarity. By default, the section will inherit the display style of the view from which it was made.

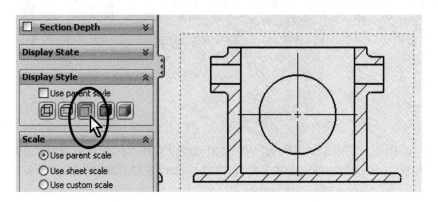

191

205. - The next step is to add a "**Detail View**". From the "View Layout" tab in the Command Manager, select the "Detail View" icon. Similar to a "**Section View**", the "Detail View" activates the "Circle" tool to draw the detail area.

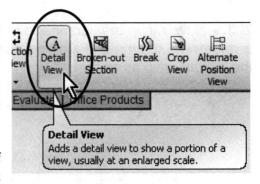

Detail View
Adds a detail view to show a portion of a view, usually at an enlarged scale.

Draw a circle in the upper left area of the "Section" view. Try not to add automatic relations when drawing the circle…

 To draw sketch elements without automatically adding geometric relations hold down the "Ctrl" key while drawing. This technique works in the part, assembly and detailing environments.

206. - …and just like the "Section" view, move the mouse to locate the Detail above the section using the dynamic preview. By default, detail views are two times bigger than the view they came from. This option can be changed in the menu "**Tools, Options**"; under the "System Options" tab, select the "Drawings" section and change the "**Detail view scaling**" factor to the number of times bigger than the view it came from.

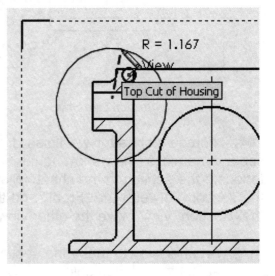

 If the detail is not as big as needed, click and drag the detail's circle and/or its center to resize and move the detail area; the "Detail" view will update dynamically, that is the reason why we don't want to add any geometric relations when drawing the circle.

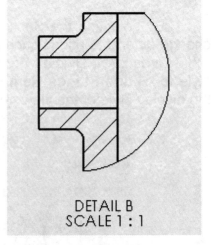

DETAIL B
SCALE 1 : 1

 Just as with the Section View, we can draw the circle (or any closed contour like ellipse, polygon or spline) and then select the "Detail View" icon.

207. - Now that we have all the views that we need in the drawing, the next step is to import the Housing's dimensions from the part into the drawing. If you remember, we added all the necessary dimensions to the part when it was modeled, and now we can import those dimensions over to the drawing, effectively reducing the amount of work needed to complete this task.

Select "Model Items" from the "Annotation" tab in the Command Manager or go to the menu "**Insert, Model Items**". Both will display the dialog to import dimensions and annotations from the model.

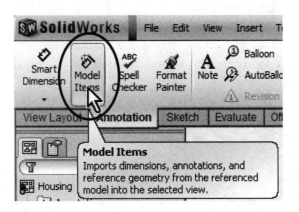

208. - In the "Model Items" options select the type of dimensions and annotations we want to import to the drawing. For this exercise select the options listed below. Remember to select "Source: Entire Model" to import the dimensions of the entire Housing, and activate the checkbox "**Import items into all views**" to add all dimensions to all views. "Dimensions Marked for Drawing" is selected by default, optionally we can activate "Hole Wizard Locations" to import the location of holes made with the Hole Wizard, and "Hole Callout" to add the machining annotations to the holes. Since we don't have any here, we'll use them later.

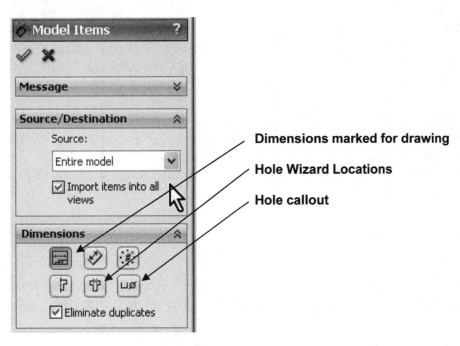

Dimensions marked for drawing

Hole Wizard Locations

Hole callout

209. - After selecting OK, SolidWorks imports the dimensions and annotations to all views, arranging them automatically. Dimensions are added first to Detail views, then to Section views, and finally main views. While SolidWorks makes a good job at adding dimensions to views, they are not always added to the view that best displays it, and here is where we have to do some work.

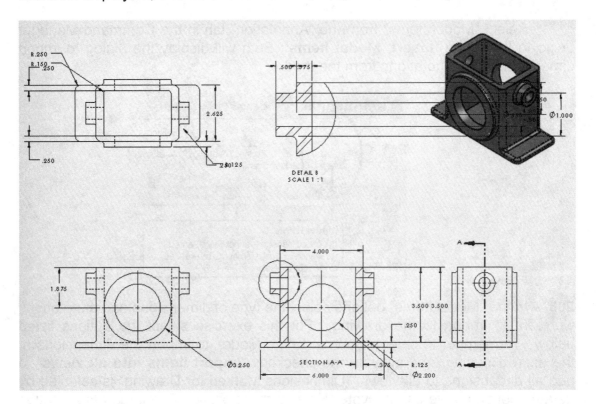

210. - The only thing for us to do now is to arrange the dimensions in order to make them easier to read. To move dimensions and annotations in the drawing simply Click and Drag them as needed. To move a dimension from one view to another, hold down the "Shift" key, <u>and while holding it down</u>, drag the dimension to the view where we want it. Notice that the dimension has to be dragged inside the view border to move. Arrange the annotations as needed to make the drawing easy to read.

 Some of the dimensions added to the Detail view are incorrectly assigned to it, so we have to move them to a different view. A good way to tell which dimensions belong to a view is simply to move the view a little. All of the dimensions attached to it will move with it and you'll be able to put them where you need them.

211. - If missing, add the center marks using the **"Center Mark"** icon from the Annotate Toolbar. Select "**Center Mark**" from the "Annotation" tab, and click on the circular edges to add center marks as needed. Click OK to finish. "**Centerlines**" can be added the same way by clicking on the cylindrical faces.

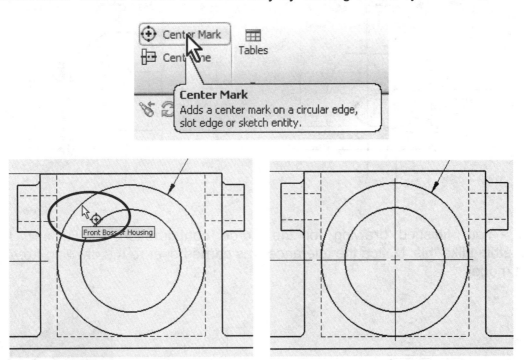

212. - Another thing that you may want to do is to reverse the arrows of a dimension. To do this, simply select the dimension and click in the green dots in the arrow heads to reverse them. To delete a dimension that is duplicated or you don't want, select it, and hit the "Delete" key in your keyboard. You only delete it from the drawing, but not from the part. If you re-import the dimensions to the drawing, SolidWorks will only bring back the ones that are missing. We'll cover more detailing and annotations in the following drawings.

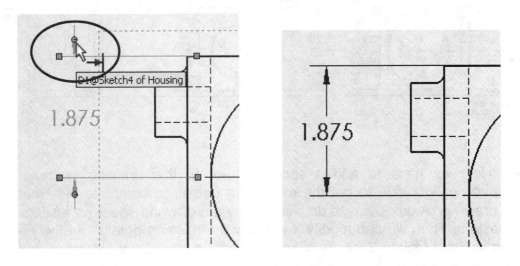

Extension lines from the dimensions can be dragged by their endpoints to change their location.

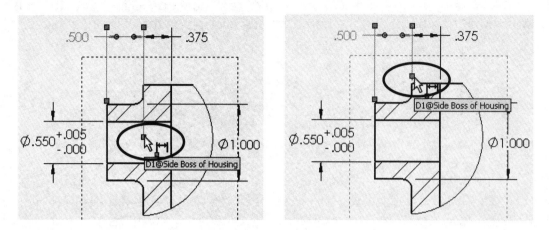

213. - Your finished drawing for the Forge configuration should now look something like this. Notice the tolerance was carried over to the drawing from the part model.

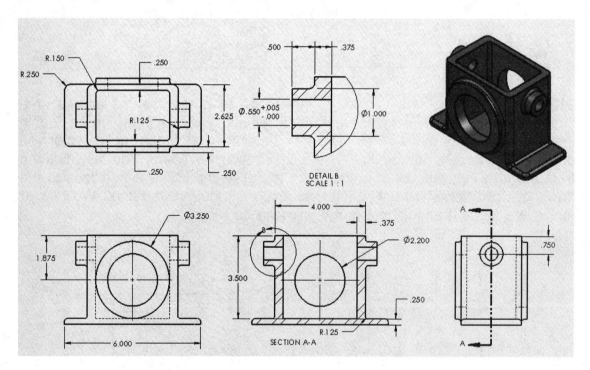

214. - Now we have to add a second sheet to this drawing, and add the Machined configuration to get the machining details to complete the Housing's detail drawing. A quick way to do it is to copy the drawing sheet we just finished, and paste it. Then we can modify it to show the missing details. In the Feature Manager Right-Mouse-Click in "Sheet1" and select "Copy", then repeat and select "Paste". Accept the defaults to continue.

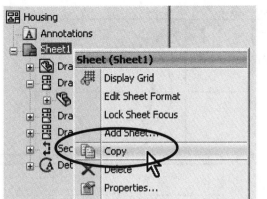

215. - Now we have a second sheet in the same drawing exactly the same as the first one. We need to tell SolidWorks to show the Machined configuration in this sheet. Make sure we are in the second sheet by clicking in the second tab on the lower left corner of the screen.

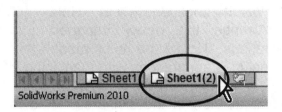

 Drawing sheets can be renamed by Right-Mouse-Clicking the sheet's tab or in the Feature Manager and selecting "Properties" or "Rename".

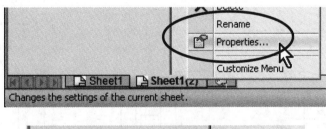

Rename "Sheet1" to "Forge" and "Sheet1(2)" to "Machined".

216. - Select the "Machined" sheet to activate it. Since we made a copy of the "Forge" drawing, we need to change the displayed configuration to "Machined". Select the Front view, and from its properties, select the "Reference Configuration" drop-down menu and change to "Machined". Notice that the view is automatically updated in the graphics area. Repeat this process for the rest of the drawing Views. The Section and Detail views will be updated after their parent views are updated.

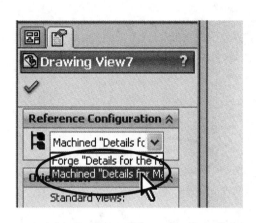

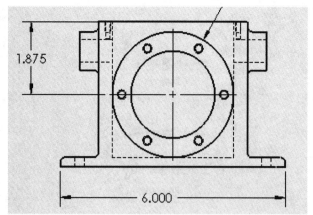

217. - Once we have changed all the views to the Machined configuration, we can repeat the "Model Items" command to import the missing dimensions and annotations in this configuration. This time include the "Hole Wizard Locations" and "Hole Callout" buttons. Arrange the newly imported dimensions and annotations. Notice the holes made

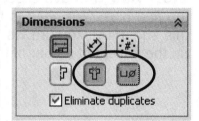

with the Hole Wizard includes annotations with machining information.

218. - Add a second detail view in the slot area. Add the missing dimensions using the "Smart Dimension" tool from the "Sketch" tab in the Command Manager as well as the missing "**Centermarks**".

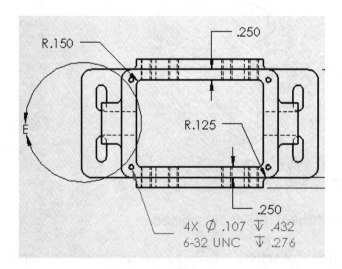

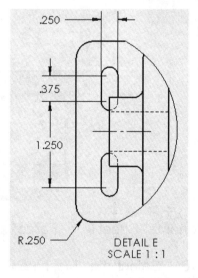

Rearrange the views as needed to make room for dimensions and annotations.

 In the "Machined" drawing sheet there are many dimensions duplicated from the "Forge" configuration. These dimensions are usually deleted to make drawings less cluttered and easier to read.

198

Forge configuration finished drawing sheet.

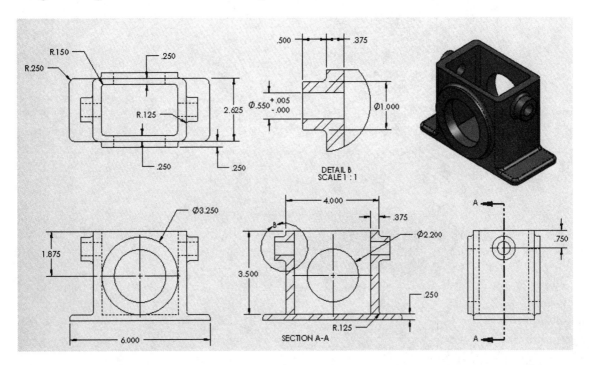

Machined configuration finished drawing sheet.

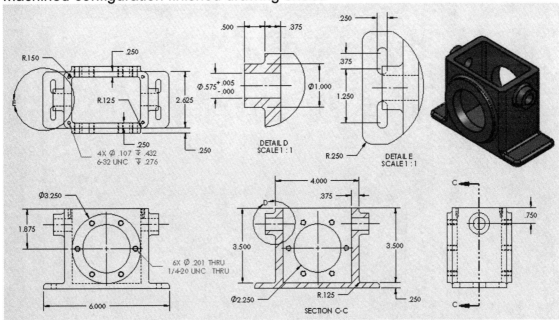

219. - Save the finished drawing as "Housing" and close the file. Note the file extension for drawings is *.slddrw.

Notes:

Drawing Two: The Side Cover

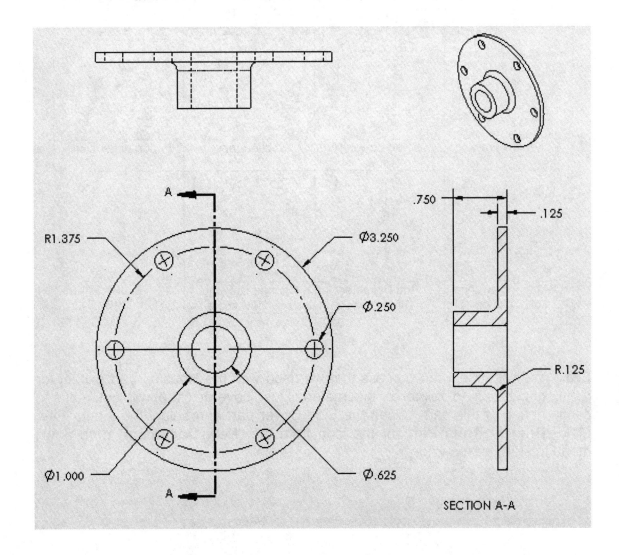

SECTION A-A

In this lesson we'll review material previously covered, including how to add standard views, sections and importing annotations from the model to the drawing. We'll also learn how to add dimensions that are not present or in the desired format, and change the sheet's scale and a view's scale.

The drawing of the Side Cover part will follow the next sequence.

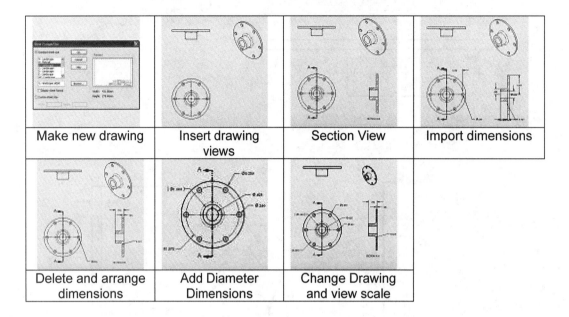

Make new drawing	Insert drawing views	Section View	Import dimensions

Delete and arrange dimensions	Add Diameter Dimensions	Change Drawing and view scale

220. - We will repeat the process that we used with the Housing part except for the configuration to reinforce the material just covered. We will not make a configuration of this part. Open the Side Cover part and select the menu "**File, Make Drawing from Part**" or the icon from the "**New Document**" drop down menu to start the drawing.

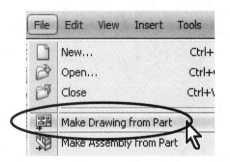

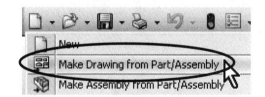

221. - For the Side Cover part we'll use the "A-Landscape" drawing template without sheet format.

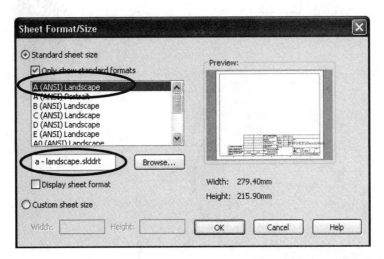

222. - From the "**View Palette**" drag and drop the "Front" view, and project the "Top" and "Isometric" views from it. Click OK when finished. Select the "Front" view and change the Display Style to "Hidden Lines Visible" as we did in the previous exercise by selecting the view and changing it in the Property Manager on the left.

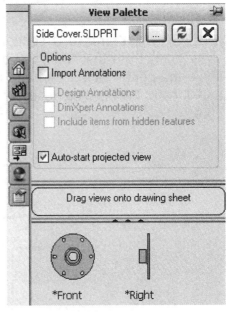

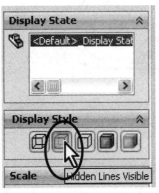

 For the "Front" and "Top" views change to "Tangent Edges Removed", and for the Isometric "Hidden Lines" and "Tangent Edges Visible".

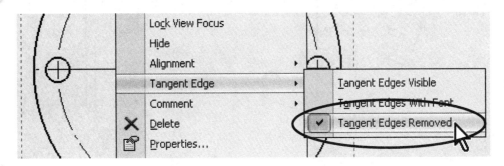

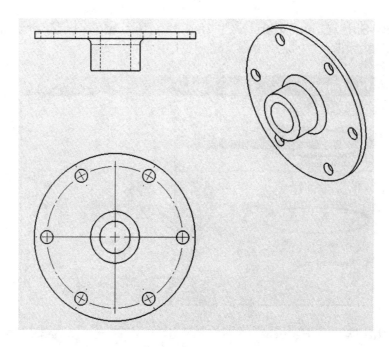

223. - Now let's make a section view of the "Front" view. Select the "Section View" icon from the "View Layout" tab; remember to *touch* the circular edge to show the center of the part and use it as a guide; draw a line through the middle and locate the section to the right. Change the section view style to "Hidden Lines"

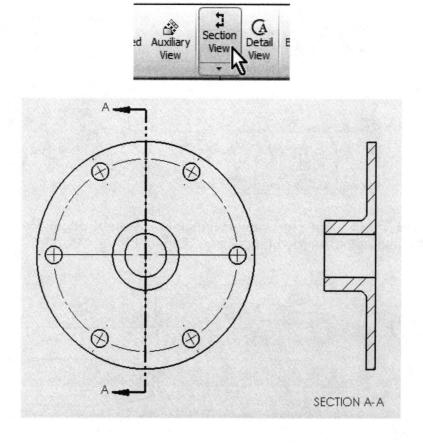

224. - Now we are ready to import the dimensions from the part into the drawing. Go to the menu "**Insert, Model Items**" or select the "**Model Items**" command from the "Annotation" tab. For "Source/Destination" select "Entire Model" and activate the checkbox "Import items into all views".

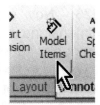

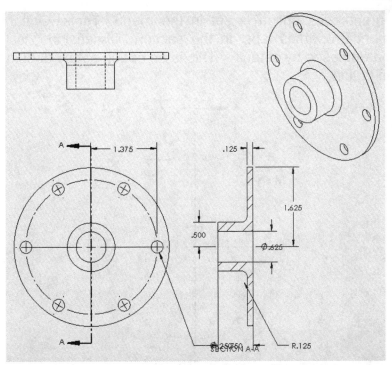

225. - When we modeled the Side Cover, we added radial dimensions in the Revolved feature, but for the detail drawing we want to show diameter dimensions instead. We will add those dimensions manually, as the imported dimensions cannot be changed to diameter the way they were done in the model. First delete the dimensions and arrange as needed to match the next image.

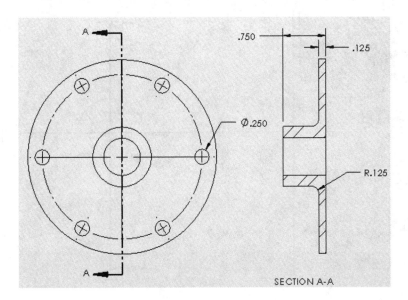

226. - Now select the "Smart Dimension" tool, and manually add the diameter dimensions indicated. Depending on the options set in the drawing template, the manually added dimensions may or may not have parentheses. Parentheses are a way to indicate the dimensions are **Reference Dimensions** that were manually added and not imported from the model.

The parentheses option is set in the menu "Tools, Options" under the Document Properties tab. In the section "Dimensions" is the checkbox "Add parentheses by default". This is an option set by document and not a system option.

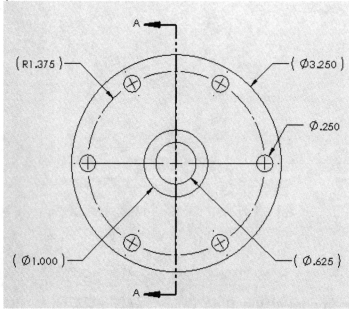

227. - To remove the **Dimension Parentheses**, Ctrl-Select the dimensions with parentheses, and from the Dimension's Property Manager, turn off the "Parentheses" option from the "Dimension Text" options box.

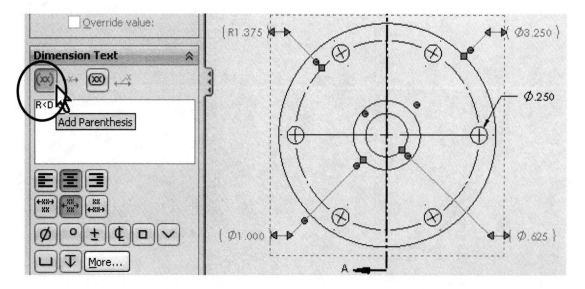

228. - Sometimes the drawing sheet is too big or too small for the views. We can change the sheet's scale to make the views fit better. In this case, we'll make the scale bigger. Make a Right-Mouse-Click *in the sheet*, not a drawing view, or on the Feature Manager on "Sheet1", and from the pop up menu select "Properties".

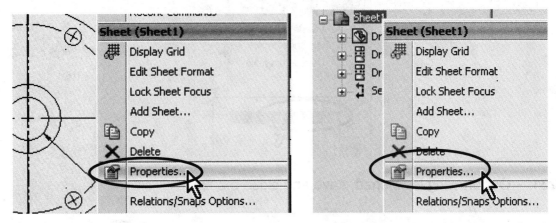

 In pop up menus notice the commands are grouped under a heading, in this case the options for "Sheet (Sheet1)" are listed.

229. - From the options box, select a scale of 1:1 and click OK. After arranging the views and dimensions, we notice that the Isometric is too big to fit nicely in our sheet, so we'll proceed to change the scale of the "Isometric" view only.

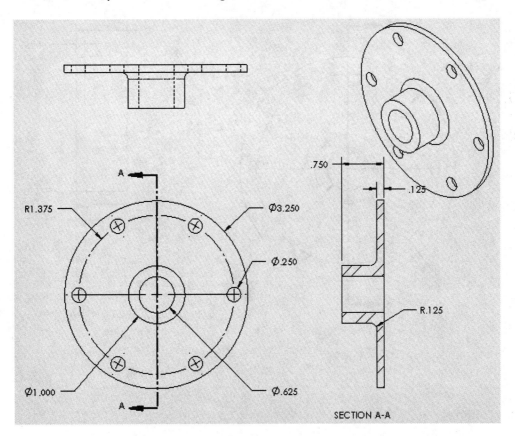

230. - Select the "Isometric" view in the graphics area, and from its Property Manager, change the view scale to 1:2 from the "**Use custom scale**" drop down menu in the "Scale" options box.

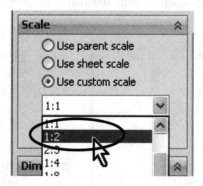

231. - Our drawing is finished, save and close the file.

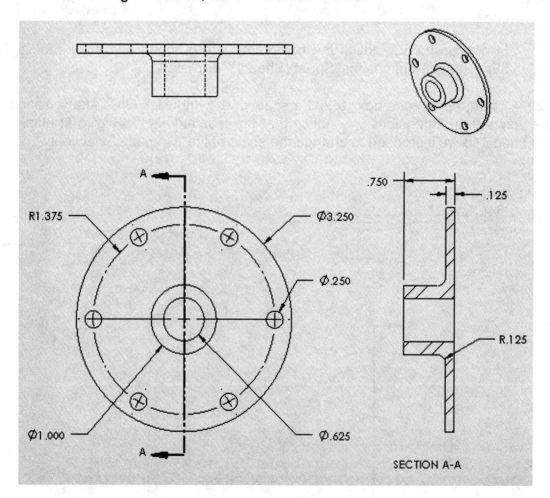

Drawing Three: The Top Cover

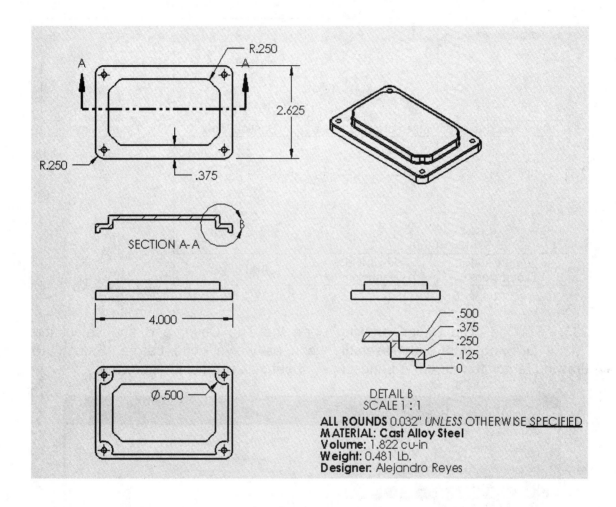

SECTION A-A

4.000

Ø.500

R.250

2.625

.375

R.250

.500
.375
.250
.125
0

DETAIL B
SCALE 1 : 1

ALL ROUNDS 0.032" *UNLESS* OTHERWISE SPECIFIED
MATERIAL: Cast Alloy Steel
Volume: 1.822 cu-in
Weight: 0.481 Lb.
Designer: Alejandro Reyes

232. - In this lesson we'll review previously covered material as well as new options like "Display only surface" in a section view, adding notes and Ordinate dimensions, as well as to add custom properties to the part and assembly files for use in drawings and bill of materials. The drawing of the Top Cover will follow the next sequence.

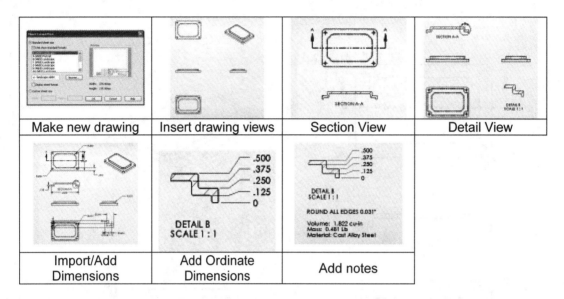

Make new drawing	Insert drawing views	Section View	Detail View
Import/Add Dimensions	Add Ordinate Dimensions	Add notes	

233. - For the "Top Cover" drawing open the Top Cover part and select the "Make Drawing from Part/Assembly" icon as we've done before. Select the drawing template with an "A-Landscape" sheet size without sheet format.

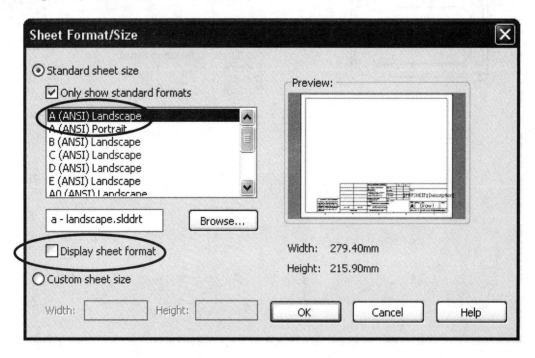

234. - Click and drag the Front view from the "View Palette" onto the sheet, and add Top, Bottom, Right and Isometric views. Change the display style to "Hidden Lines" and "Tangent Edges Removed" for all views except for the Isometric, which we'll change to "Tangent Edges Visible".

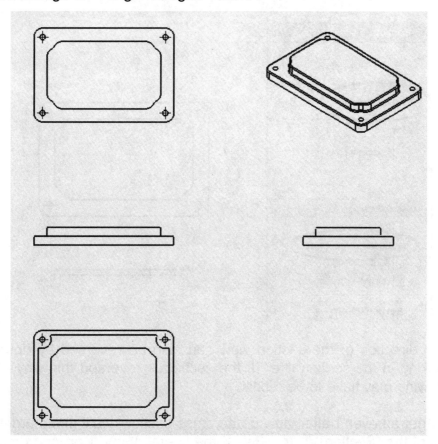

235. - Now we need to make a section through the "Top" view to get more information about the cross section of the cover (showing this view with "Hidden Lines Visible" may be confusing). Select the "Section View" icon from the "View Layout" tab and draw a horizontal line *approximately* through the middle of the "Top view" as indicated, and locate the section just below it.

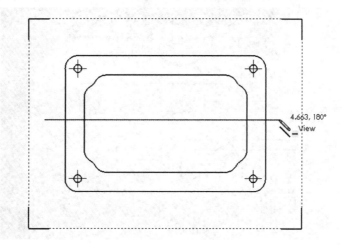

236. - After locating the section view, you may have to activate the "Flip Direction" checkbox to reverse the direction of the section view. Also check the **"Display only cut face(s)"** option to show only the section's surface ignoring the rest of the model behind the section line, as if we had only taken a thin slice of the part.

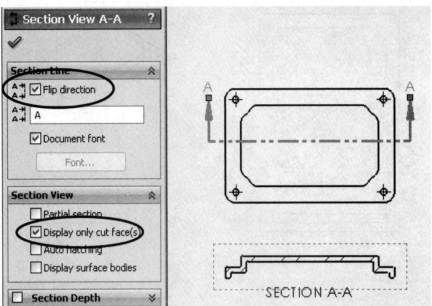

 The direction of the section view can also be reversed by double clicking in the section line. If the section is reversed this way, the drawing may have to be rebuilt.

237. - To get an even better view of the cross section, we'll make a detail of the right side of the "Section View". Select the "Detail View" icon and draw a circle as shown; locate the detail under the "Right" view to distribute the drawing evenly in the sheet. If we get Centermarks automatically added, delete them for clarity.

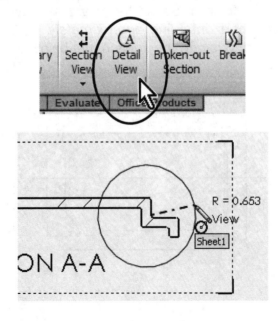

238. - When we make detail and section views, and make a mistake or we are simply not happy with the result and delete the view, SolidWorks increases the new view label to the next available letter. Then, if we make a new view, it will have this new label. The good news is that the label can be changed by selecting the view and then changing the label in the view's properties.

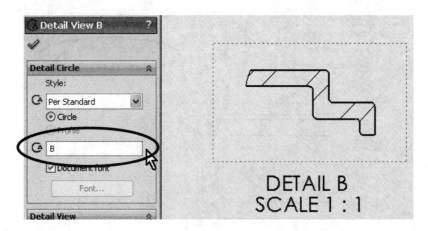

239. - The next step is to import the model dimensions from the part. Go to the menu "**Insert, Model Items**"; be sure to select the "Import items into all views" and "Entire Model" options.

Move the diameter dimension in the "Detail" view to the "Bottom" view as shown holding the "Shift" key and dragging it.

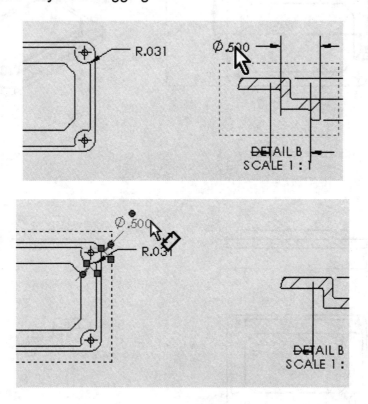

The display of the diameter dimension is not very clear this way. Make a Right-Mouse-Click in the dimension, and select "**Display Options, Display as Diameter**".

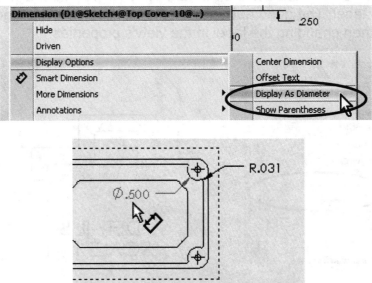

240. - Delete and arrange the rest of the dimensions as needed to match the next image.

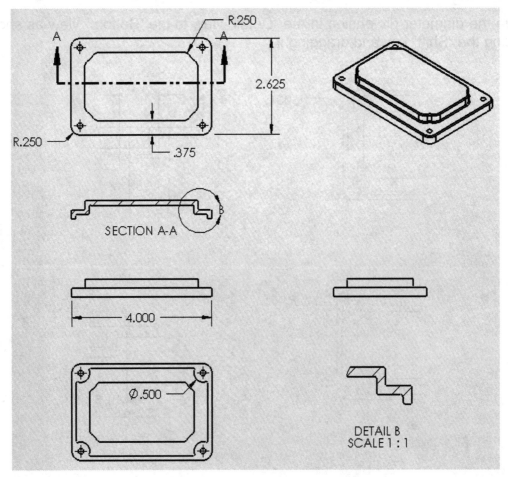

241. - Sometimes it is convenient to add "**Ordinate Dimensions**" to a drawing. The ordinate dimensions have to be added manually, unless they were added in the sketch and are being imported. Click anywhere in the graphics area with the Right Mouse Button, and from the pop up menu select "**More Dimensions, Vertical Ordinate**".

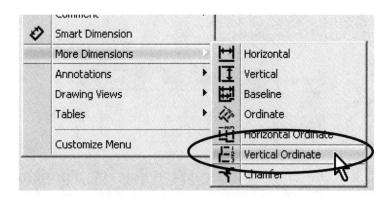

242. - Adding ordinate dimensions is very simple: First click to select the edge that will be the zero reference, click to locate the "0" dimension and finally click on each horizontal line or vertex to add a dimension to it. The Ordinate Dimensions will be automatically aligned and jogged if needed. For the Detail view, select the lower edge to be zero, click to the right to locate it, and click in the rest of the edges to add the dimensions.

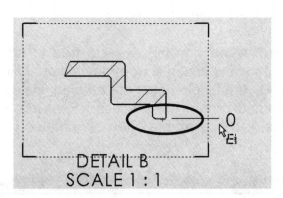

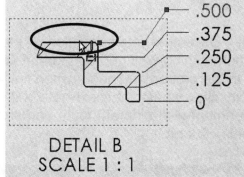

 If after adding the ordinate dimensions a dimension is missed, click with the Right Mouse Button in any of the ordinate dimensions, select "**Add to Ordinate**" and click in the edge/vertex that was missed to dimension it.

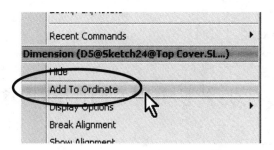

243. - The last thing we are going to do to complete the drawing is to add notes. Notes are common in drawings to communicate important information, such as material, finish, dates, etc. To add a note to our drawing, select the "**Note**" icon from the "Annotation" tab...

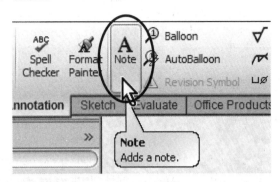

...and click in the drawing to locate the note; the **Formatting Toolbar** is automatically displayed next to the note. This toolbar is very similar to the one found in many Windows' applications to change the font and style.

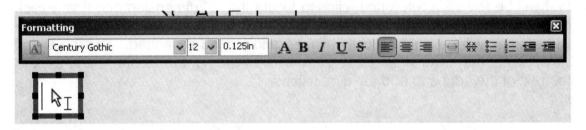

244. - Once the note's location is placed in the drawing, we are free to start typing anything we need and format just as we would in a word processor. Click OK in the property manager when finished. If multiple notes are needed, instead of OK click where the next note will be and repeat the process. Feel free to experiment with different fonts and styles. To edit a note after it is created just double click on it.

SCALE 1 : 1

ALL ROUNDS 0.032" *UNLESS* OTHERWISE SPECIFIED

245. - Parts and assemblies can be given specific "**Document Properties**" that contain user defined data like a designer's name, department, component part number, etc. or parametric information that updates if the model changes, like a component's material, weight, volume, surface area, etc. Open the "Top Cover" *part file* (Top Cover.sldprt), and select the menu "**File, Properties**". In the "Summary Information" window the "Custom" tab is selected by default. This is the area where Custom Properties are defined for a document. For this example we'll define three properties that are linked to the part's Volume, Weight and Material, and a user defined property with the designer's name.

In the "Property Name" column, click in the first cell and select "Material" from the drop down menu; in the "Type" column "Text" is automatically selected; click in the "Value/Text Expression" cell, and from the drop down menu select "Material". SolidWorks fills in the correct expression to make the property's value equal to "Cast Alloy Steel", which is the material we had assigned to this part.

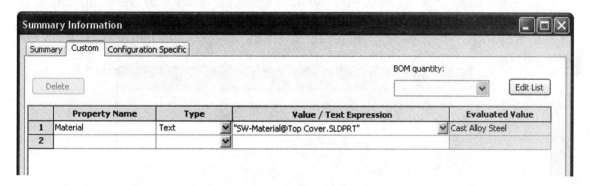

After adding a property a new empty row appears. In the second property name type "Volume" and select "Volume" for value, for the third property name select "Weight" and "Mass" for the value, and the last one type "MyName" in the "Property Name" and fill in your name for the value. It should now look like this: Click OK when done and save your changes to the Top Cover part file.

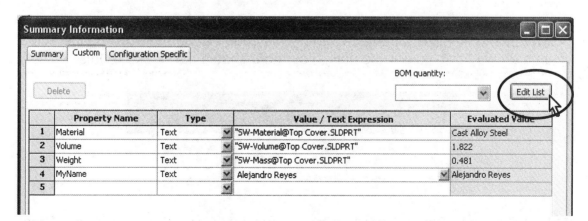

 Property names not listed can be typed in, or if it will be used in other documents select "Edit List" to add it to the drop down menu.

246. - Using the menu "**Window**" go back to the "Top Cover – Sheet1" drawing. Add a note as we just did and type: "Material:" While still editing the note click in the "**Link to Property**" icon from the note's "Text Format" options.

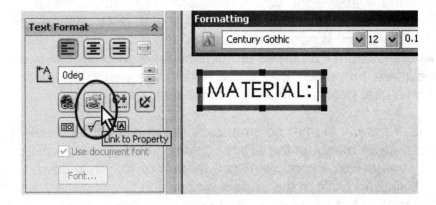

From the "Link to Property" window select "Model in view specified in sheet properties", this will tell the note to use the Custom Properties from the part in the drawing. Then select "Material" from the drop down list and click on OK when done.

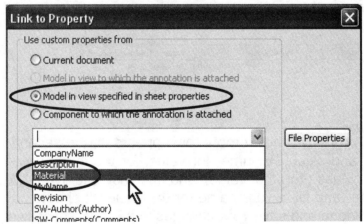

You'll see $PRPSHEET:"Material" in your note, this is the code used by SolidWorks to link to a parametric note. This code will be replaced by the actual value after we finish editing the note. Your note should now look like this:

247. - The rest of the parametric notes can be added the same way. Press "Enter" to add a new row in the note; type "Volume:" add the link to the "Volume" property; in the next row type "Weight:" and add the link to the "Weight" property; finally type "Designer:" and link to the "MyName" property.

MATERIAL: $PRPSHEET:"Material"
Volume: $PRPSHEET:"Volume"
Weight: $PRPSHEET:"Weight"
Designer: $PRPSHEET:"MyName"

MATERIAL: Cast Alloy Steel
Volume: 1.822
Weight: 0.481
Designer: Alejandro Reyes

One drawback of using parametric notes that return a numeric value is that the units are not listed, and they have to be manually added to the note or by adding a custom property to the part/assembly file with the units. The values are always displayed using the units used in the part. In our example Volume is in cubic inches and Weight is in pounds. We can add the units manually to the note. The final note will look like this:

ALL ROUNDS 0.032" *UNLESS* OTHERWISE SPECIFIED
MATERIAL: Cast Alloy Steel
Volume: 1.822 cu-in
Weight: 0.481 Lb.
Designer: Alejandro Reyes

If any of these properties change in the part, the note will be updated. For example, if the model changes size, the volume and mass values will update.

248. - Save and close the drawing file.

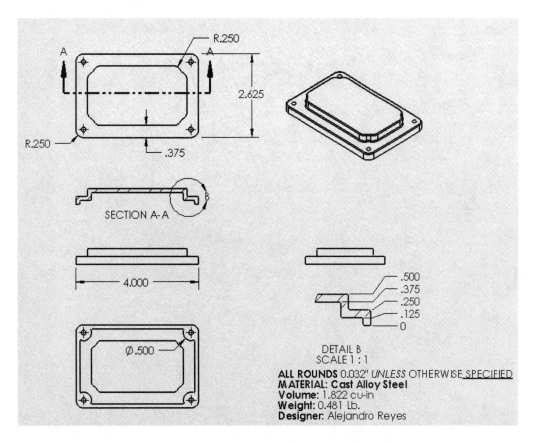

Notes:

Drawing Four: The Offset Shaft

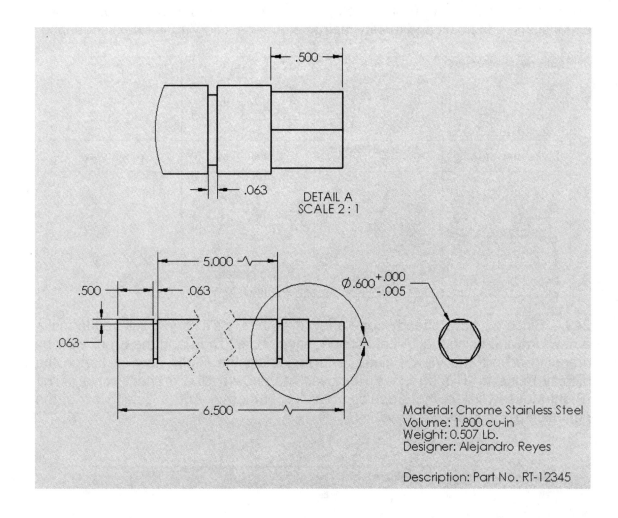

DETAIL A
SCALE 2 : 1

$\phi.600^{+.000}_{-.005}$

Material: Chrome Stainless Steel
Volume: 1.800 cu-in
Weight: 0.507 Lb.
Designer: Alejandro Reyes

Description: Part No. RT-12345

The Offset Shaft drawing, although a simple drawing will help us reinforce previously covered commands including standard and detail views, importing dimensions, manipulating and modifying a dimension's appearance, notes and custom file properties, and we'll learn how to add a break in a view. In this exercise we will make Front, Right, detail and broken views, import model dimensions, add notes and modify their appearance following the next sequence:

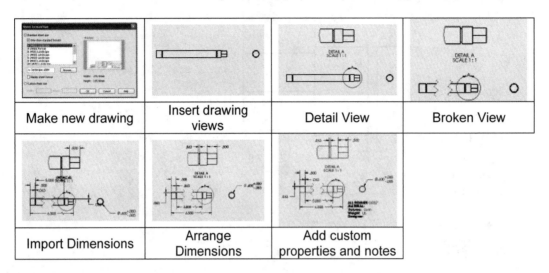

Make new drawing	Insert drawing views	Detail View	Broken View

| Import Dimensions | Arrange Dimensions | Add custom properties and notes | |

249. - Since we have already done a few drawings, we'll simply ask you to make a new drawing using the "A-Landscape" sheet size. Then add the Front view by dragging it from the View Palette, and projecting the Right view. Change the display mode to "Hidden Lines Removed" as shown. Use a sheet scale of 1:1 (Right-Mouse-Click in the graphics area, "Properties"). Add a detail view of the right side of the shaft as shown.

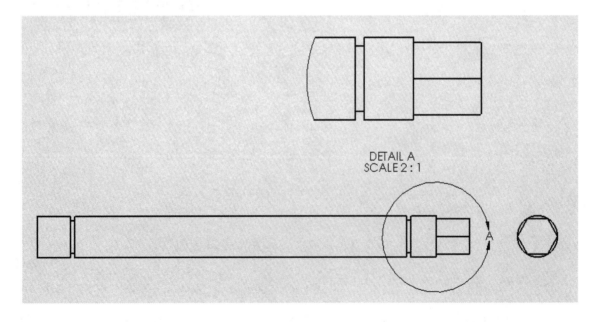

DETAIL A
SCALE 2 : 1

250. - For long elements like shafts, it's common practice to add a **Break** to shorten the view and save space in the drawing. To add a break in the front view, select the "**Break**" icon from the "View Layout" tab in the Command Manager.

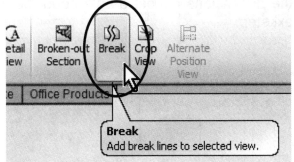

We are then presented with the "Broken View" dialog. If the view is not selected, click on it to tell SolidWorks which view to break. The Vertical Break option is already selected, and all we have to do is select where the breaks will be in the part. Click near the right side to locate the first break line, and a second click near the left side to add the second break. Click OK to finish the command.

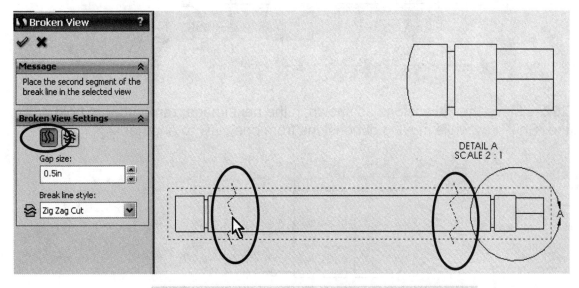

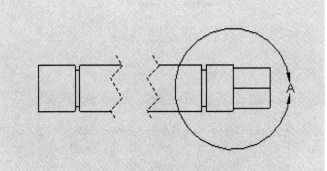

 If you need to modify the break lines location, simply click and drag them with the mouse. To delete a broken view, select a break line and delete it.

251. - You probably know what we are going to do now... Yes, import the dimensions from the part. Go to the menu **"Insert, Model Items"**; remember to select the "Import items into all views" and "Entire Model" options. In this case, SolidWorks added the dimensions almost as we need them, we'll just arrange them a little in the next step.

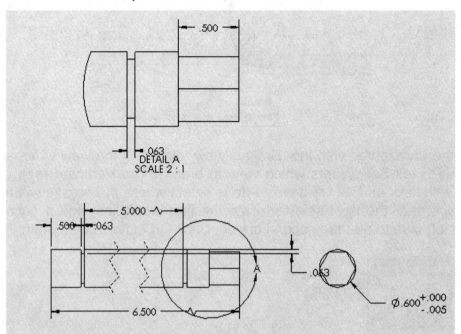

252. - Move the dimensions as shown in the next image; remember to hold down the "Shift" key while moving dimensions from one view to another.

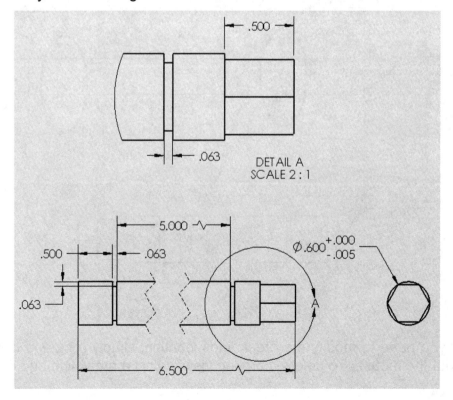

253. - Open the "Offset Shaft" part file, and from the menu "**File, Properties**" add the same custom properties as we did in the Top Cover. Add a new property called "Description" and fill in a description of the shaft. For this example anything will be OK. We'll use this description in the drawing and the bill of materials later. The "Summary Information" window should look like this:

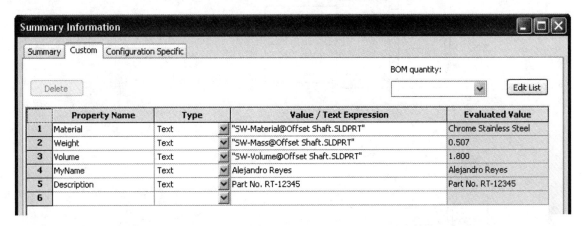

 If the values for volume and weight show as zeroes, you may have to close the file properties and rebuild the model. When you open the file properties again those values will be populated.

254. - Save the "Offset Shaft" part and go back to the drawing; add the notes using the "Note" tool from the "Annotation" tab. For this example we'll add the custom property "Description" as a separate note. The user will have to type "Material:" "Volume:" "Weight:" and "Designer:" and link to the corresponding property after each one. Remember to add the units of measure.

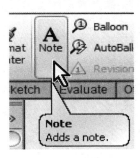

The finished note will be:

Material: Chrome Stainless Steel
Volume: 1.800 cu-in
Weight: 0.507 Lb.
Designer: Alejandro Reyes

Description: Part No. RT-12345

255. - Save and close the drawing file.

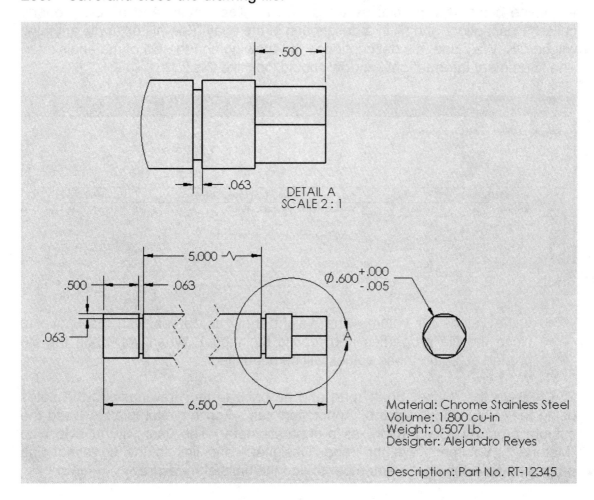

DETAIL A
SCALE 2 : 1

$\phi.600^{+.000}_{-.005}$

Material: Chrome Stainless Steel
Volume: 1.800 cu-in
Weight: 0.507 Lb.
Designer: Alejandro Reyes

Description: Part No. RT-12345

Drawing Five: The Worm Gear

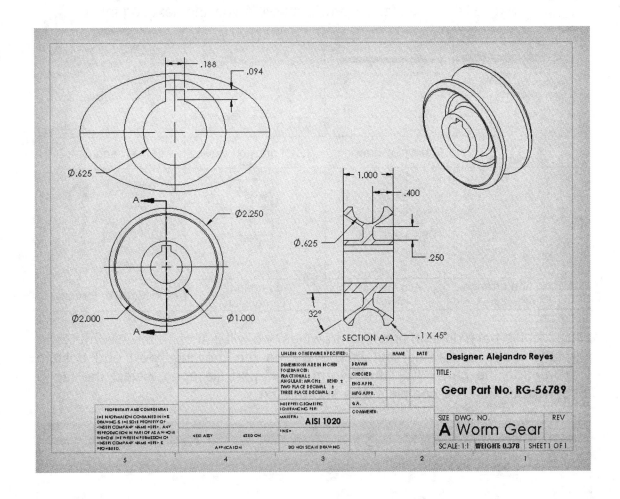

In this lesson we'll review previously covered material and a couple of new options, like adding angular dimensions, changing a dimension's precision and adding chamfer dimensions as well as a Crop View and modifying the Sheet Format. The detail drawing of the Worm Gear will follow the next sequence.

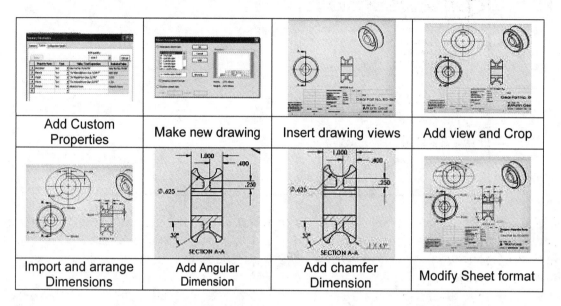

Add Custom Properties	Make new drawing	Insert drawing views	Add view and Crop
Import and arrange Dimensions	Add Angular Dimension	Add chamfer Dimension	Modify Sheet format

256. - As we have done in previous drawings, we will first open the "Worm Gear" part and make a drawing, but before we do the drawing, we'll add custom properties to the part. With the "Worm Gear" part file open, select the menu "**File, Properties**" and complete it as shown next.

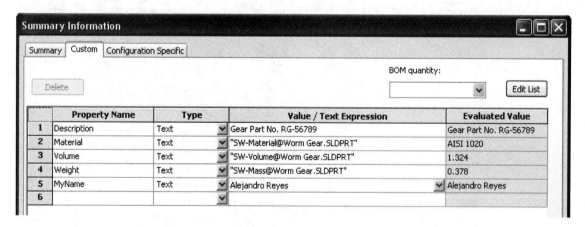

Now select the "Make Drawing from Part/Assembly" icon. For this component use the Drawing Template with an "A-Landscape" sheet size and leave the "Display Sheet Format" option active in the "Sheet Format/Size" window.

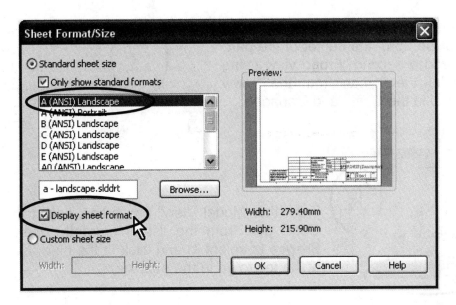

257. - Just as we've done in previous drawings, the first step is to add the main views for the model. Start by adding the "Front" view from the "View Palette" and add an "Isometric" view too. Now we need to add the section view as shown. This is a review of material previously covered. Note that we are using "Hidden Line" display mode and "Tangent Edge Removed" for the "Front" and "Section" views and "Tangent Edge Visible" for the Isometric.

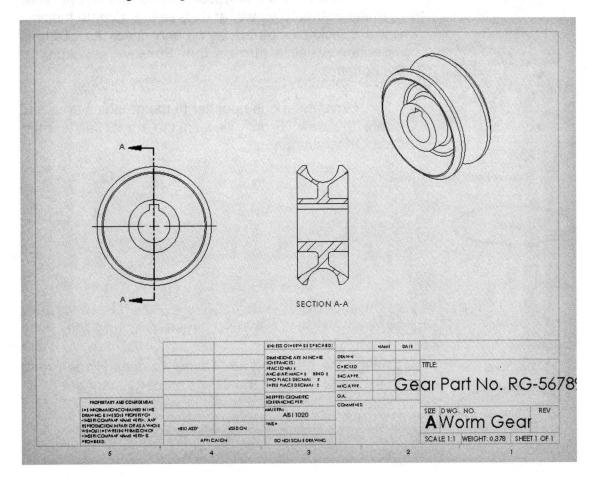

258. - After adding the "Section" view, we need a close-up of the center of the part. We will add a second "Front" view using the "Model View" icon from the "View Layout" tab in the Command Manager.

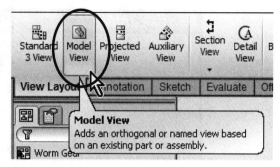

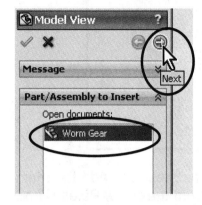

From the "Model View" properties, select the "Worm Gear" part from the list of currently open files (or Browse to select it) and click in the blue "Next" arrow at the top to go to the next step.

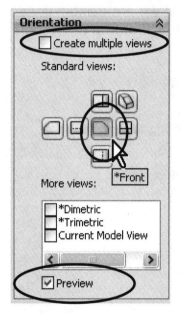

In the next set of options select the "Front" view from the Standard views, and leave the "Create multiple views" checkbox cleared, as we only need one. To help us better visualize turn on the "Preview" checkbox at the bottom.

Move the mouse pointer to the graphics area and locate the new "Front" view in the upper left corner. Click OK when done.

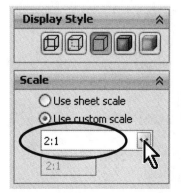

Change the new view to "Tangent Edges Hidden" and the scale to 2:1 from the view's property manager.

259. - The second "Front" view is too big, and since we are only interested in the details of its center we'll crop the view. In order to make a "**Crop View**", first we need to draw a closed profile using regular sketch tools. The closed profile can be anything including circles, rectangles, polygons, ellipses, closed splines, etc. From the "Sketch" tab in the Command Manager select the "**Ellipse**" icon. To draw an ellipse click to locate the center and then click to locate the major and minor axes. You'll see the preview as you go. Click OK when you are finished with the ellipse.

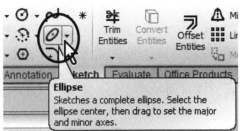

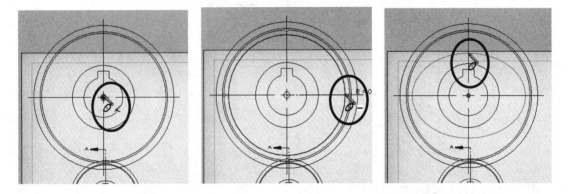

260. - With the ellipse we just drew selected (green) click in the "**Crop View**" icon from the "View Layout" tab. The view will be automatically cropped to the ellipse.

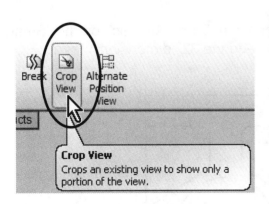

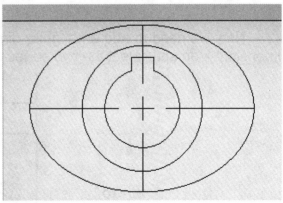

 If we need to change or delete the crop from a view, select the cropped view in the Feature Manager (it will have a scissors icon) and from the right mouse button menu select "**Crop View, Edit Crop**" or "**Remove Crop**".

261. - Now import the dimensions from the model using "Model Items". Remember to select the options "Import items into all views" and "Entire Model". After importing the dimensions, delete and arrange as needed to match the following image. You may have to change one or more dimensions to display as the diameter by right mouse clicking it and selecting "Display Options, Display as Diameter".

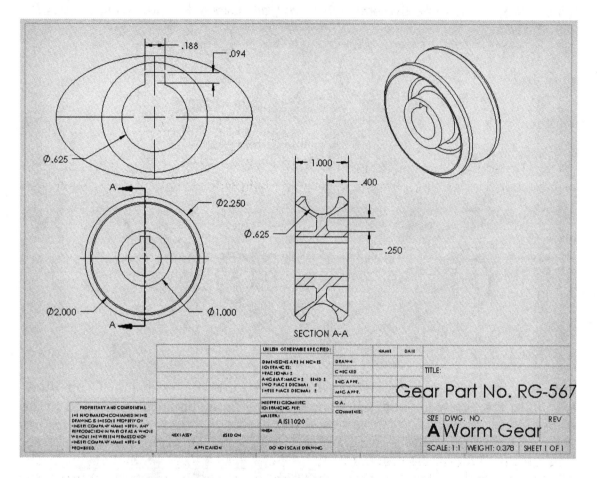

262. - Now we need to add an angular dimension. To do this click on the "Smart Dimension" icon and select the two edges to add the 32° dimension.

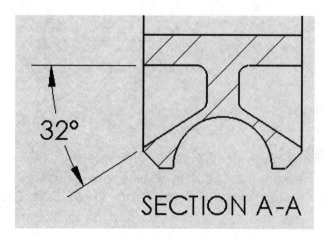

263. - The next thing we need to do is to add a "**Chamfer Dimension**". Click in the graphics area with the Right Mouse Button, and from the pop up menu select "**More Dimensions, Chamfer**" or from the menu "Tools, Dimensions, Chamfer".

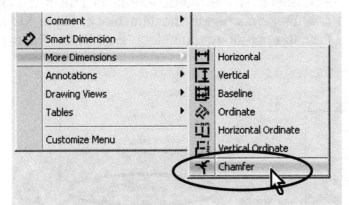

264. - To add a chamfer dimension select the chamfered edge first, then the vertical edge to measure the angle against it and finally locate the dimension.

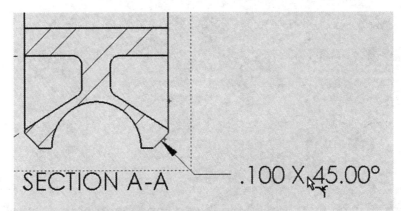

265. - Finish the Chamfer Dimension by changing its precision for distance and angle. Select the chamfer dimension, and from the dimension's properties change the "**Tolerance/Precision**" to ".1", this will change the dimension to one decimal place. For the "2nd Tolerance/Precision" select "None" to remove the decimal places from the angular value.

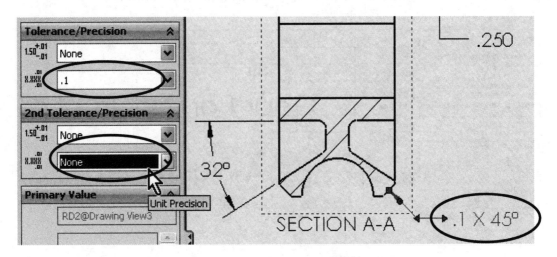

266. - In this drawing we are showing the title block. The title block is the area of the drawing where we can list all the important information; add our company name, part number, material, etc. In SolidWorks this is called the "**Sheet Format**". One characteristic of the sheet format is that it is "locked", and we cannot change it directly. To change the title block, Right Mouse Click in the drawing area and from the pop up menu select "**Edit Sheet Format**".

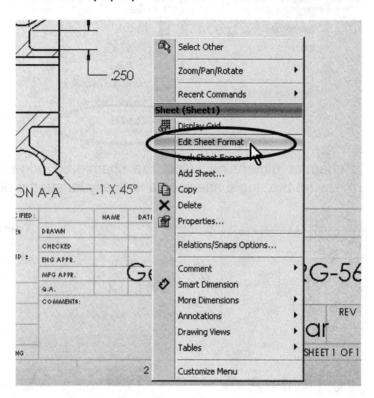

267. - We are now editing the title block and are free to change it; all the sketch tools are available to modify it as needed. Note that the drawing views are hidden when we edit the "**Sheet Format**". Some of the notes are parametric and linked to custom document properties, like the part name and material have already been filled. Other notes contain information from the drawing like the drawing's scale.

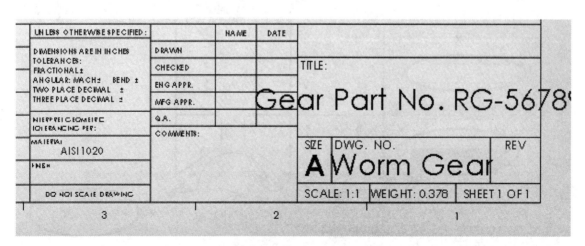

268. - Edit the notes by double-clicking them to change the font size and style as needed, and add a new note linked to your name. Note that the "Material" and "Weight" values are filled in by default, as well as the scale and the part's name. The "Title" is linked to the "Description" property. These parametric notes are from the default sheet format.

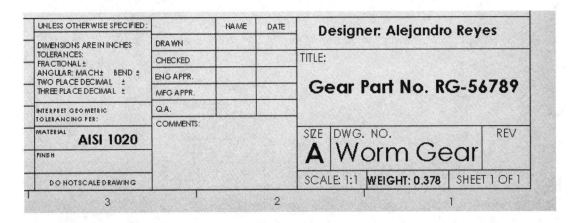

UNLESS OTHERWISE SPECIFIED:		NAME	DATE	Designer: Alejandro Reyes			
DIMENSIONS ARE IN INCHES	DRAWN			TITLE:			
TOLERANCES: FRACTIONAL± ANGULAR: MACH± BEND ± TWO PLACE DECIMAL ± THREE PLACE DECIMAL ±	CHECKED						
	ENG APPR.			Gear Part No. RG-56789			
	MFG APPR.						
INTERPRET GEOMETRIC TOLERANCING PER:	Q.A.						
	COMMENTS:						
MATERIAL AISI 1020				SIZE	DWG. NO.		REV
FINISH				A	Worm Gear		
DO NOT SCALE DRAWING				SCALE: 1:1	WEIGHT: 0.378	SHEET 1 OF 1	
3			2		1		

After modifying the title block to our liking, exit the sheet format and return to editing the drawing by selecting **"Edit Sheet"** from the right mouse button menu, or clicking in the "Sheet Format" confirmation corner.

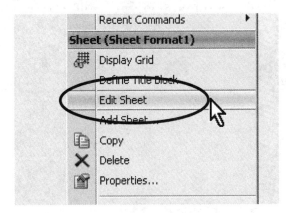

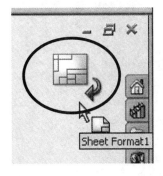

 If we modify an empty drawing's Sheet Format and save it as a template, the title block changes will be saved with the template and be available for new drawings based on this template. See the Appendix for more on templates.

269. - Save and close the drawing file.

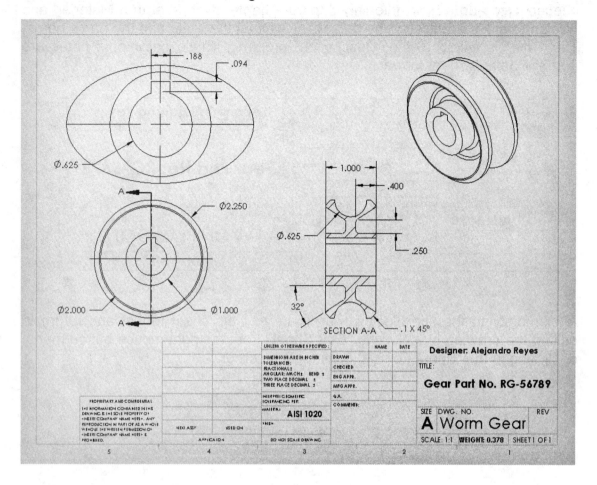

Drawing Six: The Worm Gear Shaft

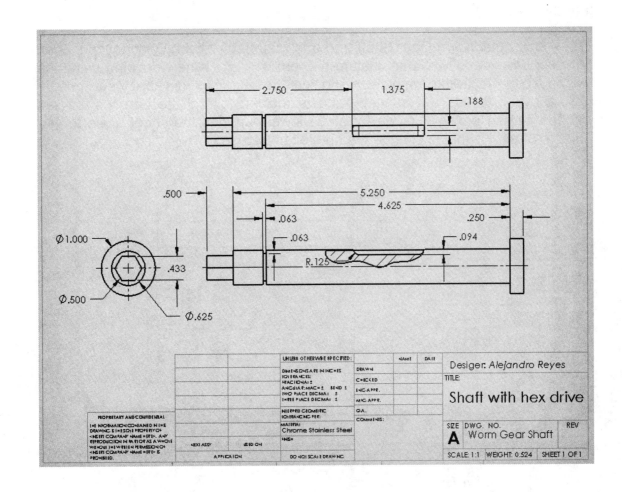

Just as we did with the Offset Shaft drawing, we'll reinforce making new drawings, adding views, importing dimensions, moving dimensions from one view to another and changing a diameter's display style. In this exercise we will make Left, Front and Top views, import and arrange the model dimensions and learn how to make a new type of view called "**Broken-Out Section**" to complete the drawing.

270. - Open the Worm Gear Shaft and add Material, Weight, Description ("*Shaft with hex drive*") and MyName custom properties. Make a new drawing using the "A-Landscape" drawing template. Add the "Front" view from the View Palette, click above it to add a Top View, and to the left to add the Left View. Use "Hidden Lines Removed" mode for all 3 views. Change the sheet scale to 1:1. Edit the Sheet Format to change the note's font size to fit.

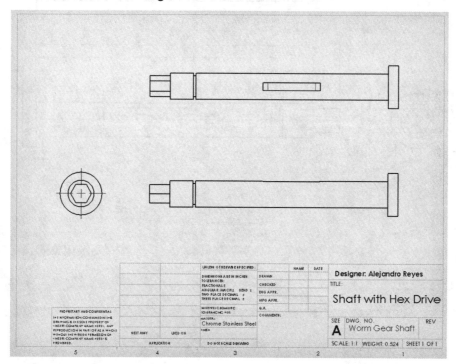

271. - Import the dimensions from the model…

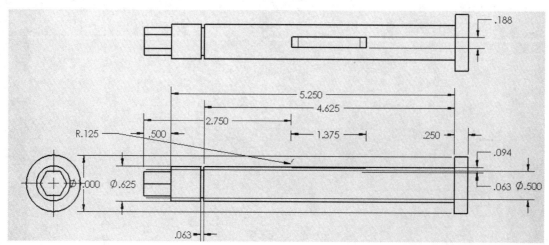

238

272. -...and arrange them as indicated.

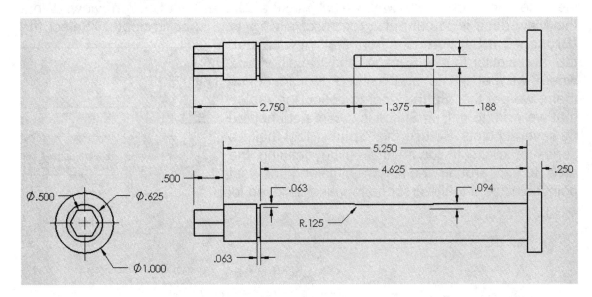

273. - One more thing we need to add to the drawing is a centerline. SolidWorks allows us to add a centerline to every cylindrical face of the model. To add centerlines to our drawing views, select the "**Centerline**" icon from the "Annotation" Tab, and click in the cylindrical surface that we need to add a centerline to. When finished adding centerlines click OK to finish.

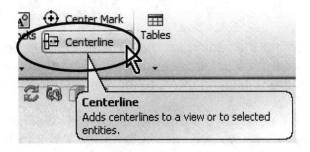

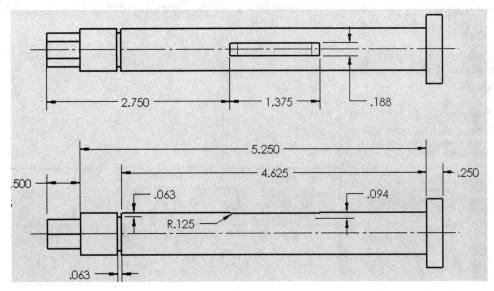

274. - Instead of changing the "Front" view display to "Hidden Lines Visible" to view the details of the keyway, we will make a "Broken-Out Section" view. What this view does is to cut-out a region of a view to a specific depth. Select the "**Broken-Out Section**" icon from the "View Layout" tab, and similar to the "Section" and "Detail" views, where we get a line and a circle tool to define them, we get the "**Spline**" tool to define the region that we want to cut. A Spline is a curve connected by several points. To use the "**Spline**" tool make a series of clicks to locate the spline defining the area to cut around the keyway making the last point coincident to the first one to make a closed loop.

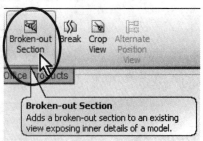

Broken-out Section
Adds a broken-out section to an existing view exposing inner details of a model.

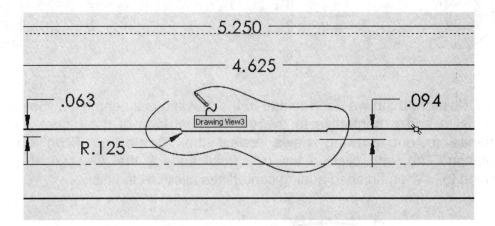

 If we want to define the region for the Broken-Out Section using a different tool, such as lines arcs or ellipses, draw a closed profile using those tools and select it before activating the "Broken-Out Section" tool.

275. - When the last point of the **Spline** is added, we get the "**Broken-Out Section**" properties, where we are asked to enter a depth to make the cut or select an edge of the model to define the depth. Select the edge indicated in the "Top" view to define how deep the cut will be.

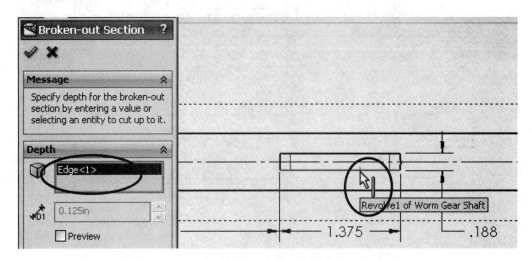

Click OK to finish

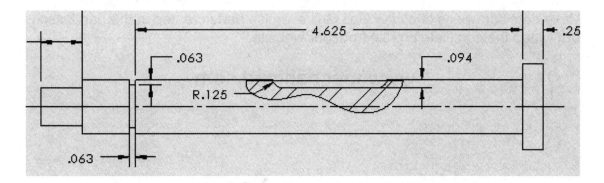

 To edit or delete the "Broken-Out Section", right mouse click <u>inside</u> the section's region and from the menu "Broken-Out Section" select "Delete", "Edit Definition" (to change its depth) or "Edit Sketch" to modify the region.

276. - In this drawing, it may be a good idea to manually add a dimension to the hexagonal cut. Save the drawing and close the file.

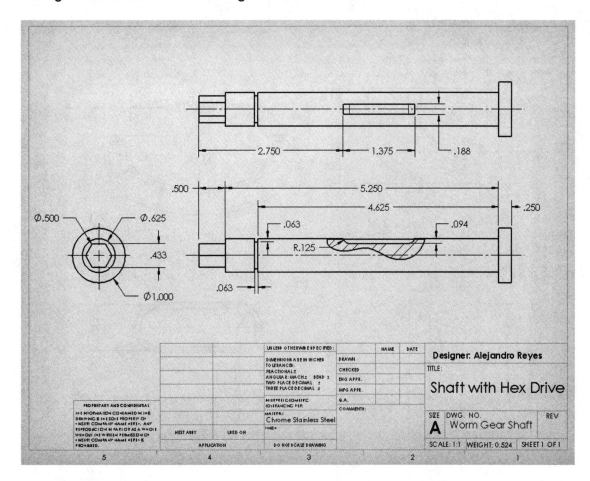

Challenge Exercises:

Make detail drawings of all the Gas Grill exercise parts created in the part lesson. Download the complete project from our website.

www.mechanicad.com

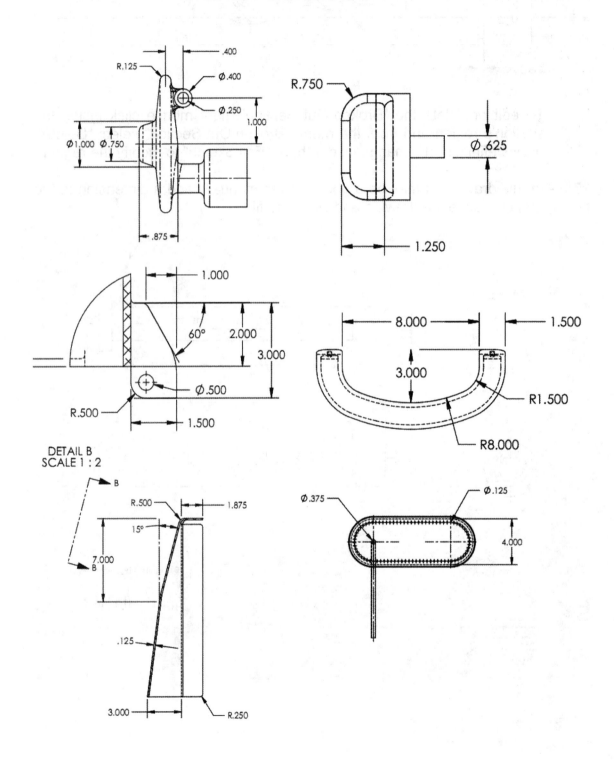

DETAIL B
SCALE 1 : 2

Assembly Modeling

The next part of the design after making the parts and drawings is to assemble all the components. Designing the parts first and then assembling them is known as "Bottom-Up Design". This is similar to buying a bicycle; when you open the box you get all the pieces needed ready for assembly. A different approach, usually known as "Top-Down Design" is where *the parts are designed while working in the assembly*. This is a very powerful tool that allows us to match parts to each other, changing a part if another one is modified. In this book we'll cover the Bottom-Up Design technique, since it is easier to understand and is also the basis for the advanced Top-Down Design. In general, it's a good idea to design the parts, make the full assembly to make sure everything fits and works as expected (Form, Fit and Function), *and then* make the drawings of the parts and assemblies; this way the drawings are done at the end, when you are sure everything works correctly.

So far, we've been working on parts, single components that are the building blocks of an assembly. In the assembly we have multiple parts, and maybe other assemblies (in this case called sub-assemblies). The way we tell SolidWorks how to relate components (parts and/or sub-assemblies) together is by using Mates or relations between them. Mates in the assembly are similar to the geometric relations in the sketch, but in the assembly we reference faces, planes, edges, axes, vertices and even sketch geometry from the features in the components.

In an assembly, the components are added one at a time until we complete the design. In a SolidWorks assembly every component has six degrees of freedom, meaning that they can move and rotate six different ways: three translations in X, Y and Z, and three rotations about the X, Y and Z axes. Once one component is mated to another component, we restrict how it moves, or which degrees of freedom are constrained. This is the basis for assembly motion and simulation.

The first component added to the assembly has all six degrees of freedom fixed by default. Therefore, it's a good idea to make the first component the one that will serve as a reference for the rest of the components. For example, if we make a bicycle assembly, the first component added to the assembly would be the frame. For the gear box we are designing, the first component added to the assembly will be the Housing, as the rest of the components will be attached to it.

Notes:

The Gear Box Assembly

In making the Gear Housing assembly we will learn many assembly tools and operations, including making new assemblies, how to add components to it, add Mating relations between them using model faces and planes, adding fasteners, cover part design tables, interference detection and exploded views of the assembly and how to change a part's dimension in the assembly. The sequence to follow to complete the Gear Housing assembly is the following:

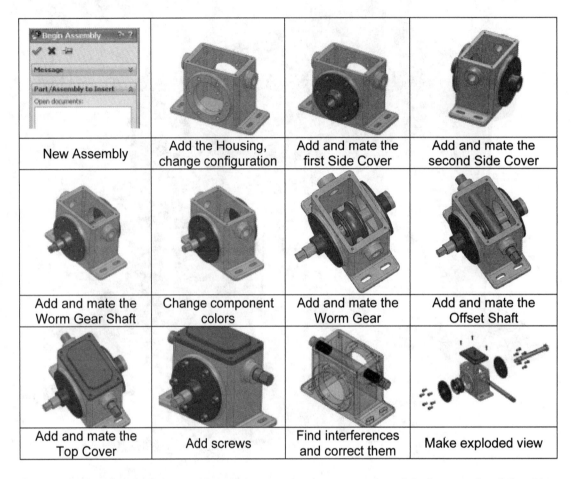

New Assembly	Add the Housing, change configuration	Add and mate the first Side Cover	Add and mate the second Side Cover
Add and mate the Worm Gear Shaft	Change component colors	Add and mate the Worm Gear	Add and mate the Offset Shaft
Add and mate the Top Cover	Add screws	Find interferences and correct them	Make exploded view

277. - The first thing we need to do to make a new assembly is to select the New Document icon, select the Assembly template and click on OK.

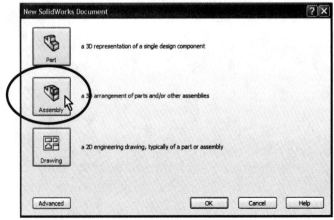

278. - The first thing that we see when we make the new assembly is the "**Begin Assembly**" dialog to start adding components. As we discussed previously, the Housing will be the first component added. Click on "Browse" to locate the Housing part file, select it in the open dialog box making sure we are using the "Machined" configuration in the "Configurations" drop-down menu. Click "Open" when done. If you have the "Graphics Preview" option box checked, you will see a preview of the component being inserted. If you cannot see the component that you want, make sure you are looking in the correct folder and have the "Files of Type" set to "Part" (it can be either "Part" or "Assembly").

279. - We now have to locate the Housing in the assembly; as you can see, the preview follows the mouse. What we want to do is to locate the Housing at the assembly origin. If you cannot see the assembly origin, turn it on by selecting the menu "**View, Origins**" (this can be done while you are inserting a component). The reason to locate the Housing at the assembly's origin is to have the Housing's planes and origin aligned with the assembly's planes and origin. To add the housing AND align it with the

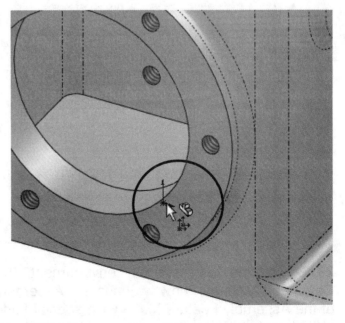

Assembly's origin, simply click the OK button, or move the mouse pointer to the assembly origin and click on it. (The cursor will have a double arrow next to it and you will see the housing "snap" in place before clicking.)

280. - If the Housing is loaded using the "Forge" configuration, we can change it by selecting it in the Feature Manager and clicking the "Component Properties" icon from the pop up toolbar.

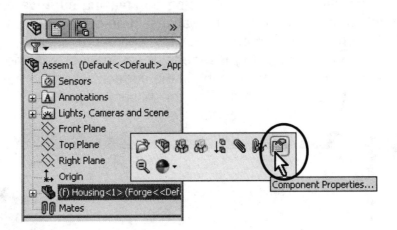

 If the component added to the assembly has configurations, the selected configuration is shown next to the part's name in parenthesis in the Feature Manager.

From the "Referenced Configuration" box select the "Machined" configuration and click OK. The Housing in the graphics area should now show the Machined configuration. When we added the Housing to the assembly, the Forge configuration was displayed because the Housing was last saved with the Forge configuration active.

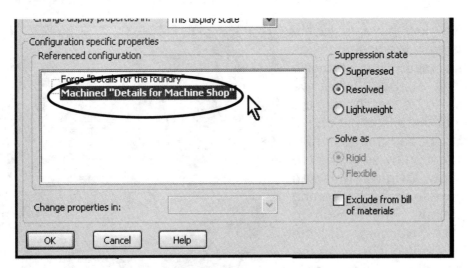

281. - Once in the Assembly environment, the toolbars are changed in the Command Manager; now we have an "Assembly" tab. Also notice at the bottom of the Assembly Feature Manager a special folder called "Mates". This is where the relations between components (Mates) will be stored.

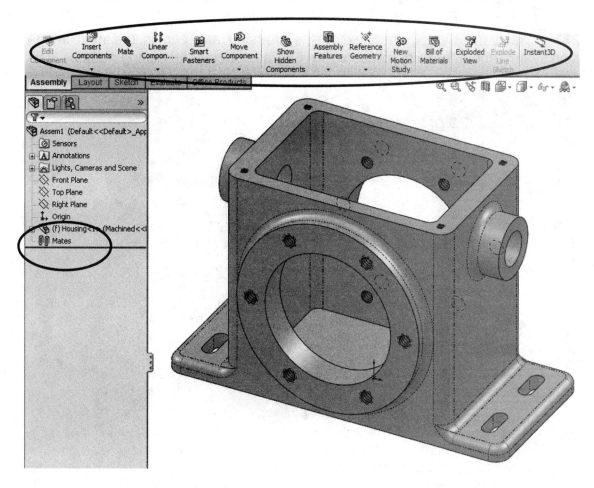

282. - To the left of the "Housing" in the Feature Manager, we can see a letter "**f**"; this means that the part is "Fixed" and its six degrees of freedom are constrained. This is the behavior when we insert the first part, or when we click OK in the Insert Component dialog instead of clicking in the graphics area. To add a second component to our assembly click the "**Insert Components**" icon...

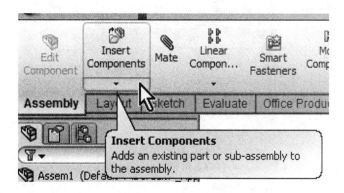

...browse to the folder where "Side Cover" was saved and open it.

283. - Place the Side Cover next to the Housing as seen in the next image. Don't worry about the exact location; we'll locate it accurately using mates.

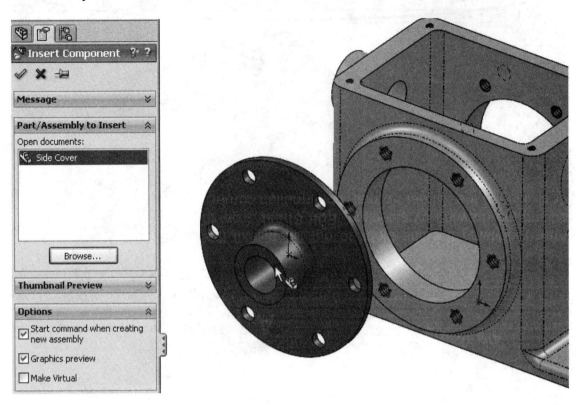

284. - The "Side Cover" part name in the Feature Manager is preceded by a (-), this means that the part has <u>at least </u>one unconstrained degree of freedom. Since this part was just inserted, all six degrees of freedom are unconstrained and the part is free to move in any direction and rotate about all three axes.

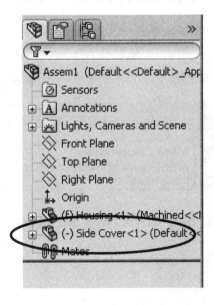

285. - Now we are ready to add the relations (mates) between the "Housing" and the "Side Cover". The mates will reference one component to another restricting their motion. As explained earlier, mates can be added between faces, planes, edges, vertices, axes and even sketch geometry. Click in the "**Mate**" icon or select the menu "**Insert, Mate**" to reference the "Side Cover" to the "Housing".

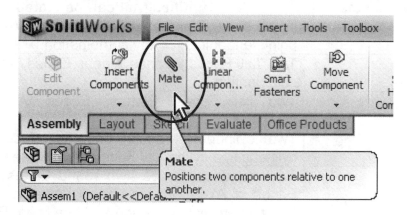

 When possible, select model faces as your first option; they are easier to select and visualize most of the time.

286. - For the first mate, select the two <u>cylindrical faces</u> indicated in the next picture. As soon as the second face is selected, SolidWorks recognizes that both faces are cylindrical and automatically "snaps" them with a **Concentric Mate**, (SolidWorks defaults to concentric as it is the most logical option). The "Side Cover" is the one that moves because it is the part with unconstrained degrees of freedom; remember the "Housing" was fixed when it was inserted.

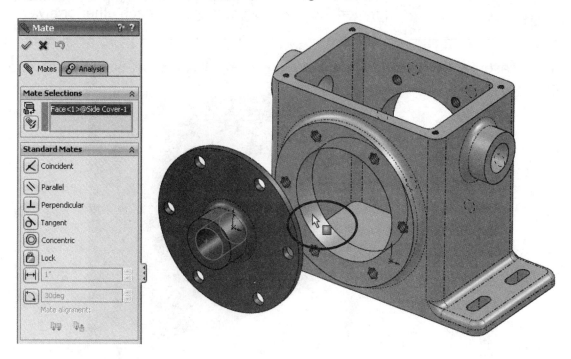

287. - As soon as we select the second face, a pop up toolbar shows options to mate the entities. (These options are listed in the Property Manager too.) This toolbar helps us be more productive and minimize mouse travel. The following table shows basic mate options available

Standard Mates	Entities that can be mated
⚔ Coincident	Two Faces, Planes, Edges, Vertices, Axes, Sketch points/endpoints, or any combination.
⟍ Parallel	Two flat Faces, Planes, linear Edges, Axes or any combination.
⊥ Perpendicular	Two flat Faces, Planes, Edges, Axes or any combination.
⤢ Tangent	Two cylindrical Faces, one flat Face and one cylindrical, a cylindrical Face and one linear Edge.
◎ Concentric	Two cylindrical Faces, two round Edges, two linear Edges or Axes, one cylindrical Face and one round Edge, one cylindrical Face and one linear Edge.
🔒 Lock	This option constrains all degrees of freedom of the component locking it in place.
⊢ 1" ☑ Flip dimension	Specify a distance between any two valid entities for Coincident or Parallel mate. "Flip dimension" reverses the direction to one side or the other. Mating two flat Faces or Planes will also make them parallel.
☐ Flip dimension 0deg	Specify an angle between any two valid entities for Coincident, Parallel or Concentric mates. "Flip dimension" will reverse the direction of the angular dimension.
Mate alignment: ⊞⊟ ⊞⊞ or ↗↙	Reverses the orientation of the components being mated. For two Faces, to look at each other or away.

In this case the **Concentric Mate** is pre-selected; click the OK button in the pop up toolbar to add this mate.

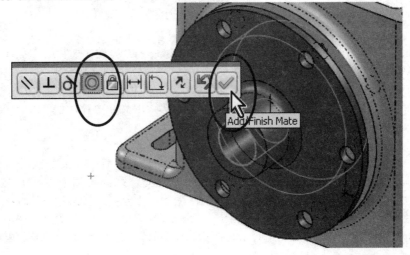

288. - Notice that after adding the first mate the dialog remains visible and ready for us to add more mates. It will remain active until we click **Cancel** or hit "Esc" on the keyboard. Notice that the mate added is listed under the "Mates" box in the Property Manager.

289. - For the second mate, select the two cylindrical faces indicated (one from the "Side Cover", and one from the "Housing"); SolidWorks defaults again to a **Concentric** mate and rotates the "Side Cover" to align the holes. This will prevent the "Side Cover" from rotating. We can use the "**Magnifying Glass**" to make selection of small faces easier. Click OK to add the mate.

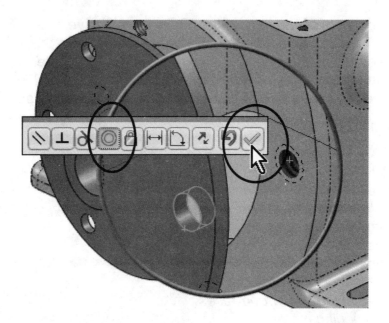

In this case the concentric mate works as expected because the holes are located <u>exactly</u> at the same distance from the center in both parts; in reality, it is generally a better idea to align the two components using planes and/or faces, with either a parallel or coincident mate, as will be shown later in the book.

290. - The last mate will be a **Coincident Mate** between the back face of the "Side Cover" and the front face of the "Housing"; rotate the view if needed to select the faces. The cover will move to match the faces coincident and **Coincident Mate** will be pre-selected. Click OK on to add the mate and finish the command (if the Side Cover is *inside* the Housing, click and drag to move it).

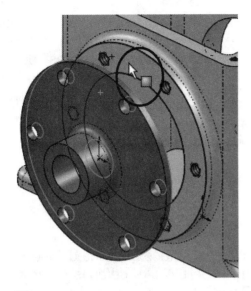
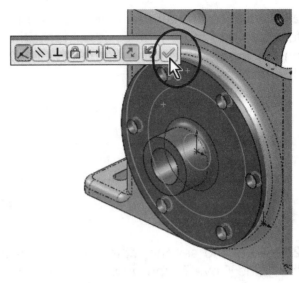

 We can also use the "**Select Other**" tool to select hidden faces.

291. - All six degrees of freedom of the "Side Cover" have now been constrained using mates; this can be seen in the Feature Manager where the "Side Cover" is no longer preceded by a (-) sign. Notice that the "Mates" folder now includes the two Concentric and one Coincident mate we just added; SolidWorks automatically adds the names of the components related in each mate. Note we can change the width of the Feature Manager by dragging the right side of it to make it wider.

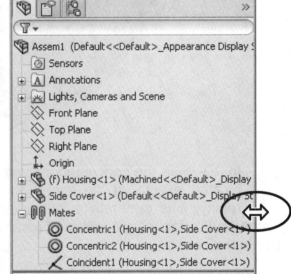

*IMPORTANT: If a (+) sign precedes a part name, you probably also received an error message telling you that the assembly had been **Over Defined**. If this is the case, it means you added conflicting mates that cannot be solved, or inadvertently selected the wrong faces or edges when adding mates.*

The easiest way to correct this error is to either hit the Undo button

or delete the last mate in the "Mates" folder; you will be able to identify the conflicting mate because it will have an error icon next to it. If multiple mates have errors, start deleting the last mate at the bottom (this is the last one added); chances are this is the cause of the problem. If you still have errors, keep deleting mates with errors from the bottom up until we clear all the errors. It's not a good idea to proceed with errors, as it will only get worse.

292. - We are now ready to insert the second "Side Cover". Repeat the "**Insert Component**" command to add a second "Side Cover" and locate it on the other side of the Housing approximately as shown. Don't worry too much about the exact location; we'll move and rotate the part in the next step.

 A quick way to add a copy of a component already in the assembly is to hold down the "Ctrl" key and click-and-drag the part to be copied.

293. - When the second Side Cover is inserted it has a (-) sign next to it in the Feature Manager; remember this means that it has <u>at least</u> one unconstrained Degree of Freedom (DOF); since this part was just inserted in the assembly, all six DOF are unconstrained, and the component can be moved and rotated. To "**Move a Component**" click-and-drag it with the <u>Left Mouse Button</u>. To **Rotate** it click-and-drag with it the <u>Right Mouse Button</u>. Move and rotate the second Side Cover as needed to align it *approximately* as shown. Remember, we'll add more mates to locate it precisely.

294. - We will now add three mates to the second cover as we did to the first "Side Cover". Select the "Mate" icon and add a **Concentric** mate selecting the faces indicated next. Rotate the view (not the part) to get a better view of the faces to select. We can also select one of the faces, and from the pop-up toolbar select the "Mate" icon.

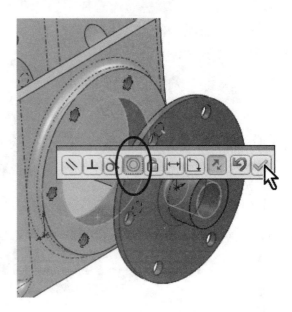

 Before adding the second Concentric mate, click and drag the cover. Notice that it can rotate about its center and also move along the axis; under-constrained components are the basis for SolidWorks to simulate motion.

295. - Now let's add the concentric mate to align the screw holes between the Cover and the Housing. Since the part is symmetrical, it doesn't really matter which two holes are selected. If the cover had a feature that needed to be aligned, then the orientation would be important.

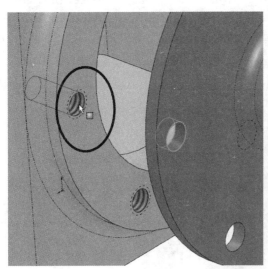
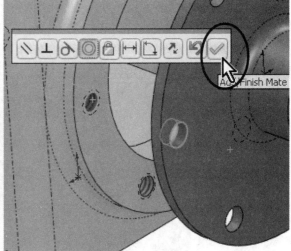

296. - Finally add a Coincident mate between the flat face of the "Housing" and the second "Side Cover" as we did with the first cover. Click OK to continue.

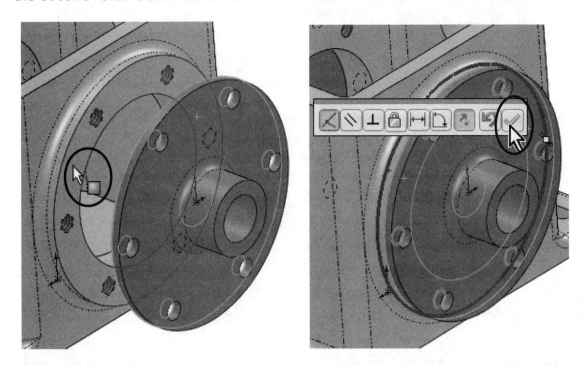

297. - It is very important to notice that these mates could have been added in any order; we chose this order to make it easier for the reader to see the effect of each mate. Whichever order you select, you will end up with both Side Covers fully defined (no free DOF) and six mates in the "Mates" folder.

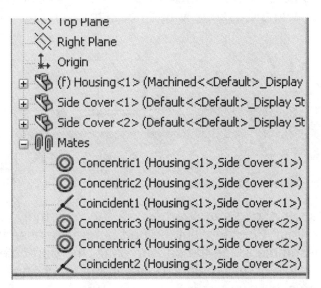

298. - Now that we have correctly mated both Side Covers, we'll add the "Worm Gear Shaft" using the same procedure as before and locating it somewhere above the "Housing".

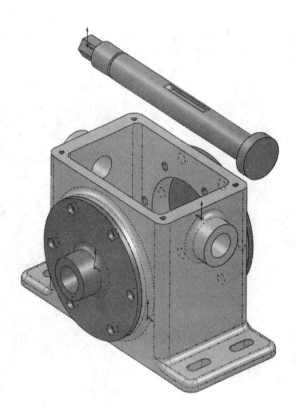

299. - With four different components in the assembly, it's a good idea to change their appearance to easily identify them. To change a component's color select the component in the Feature Manager and from the pop up toolbar select the **"Appearance"** icon. From the drop-down list select the first option, since we want to change the part's color at the assembly level only. This means that the color will change only in the assembly and not in the part file.

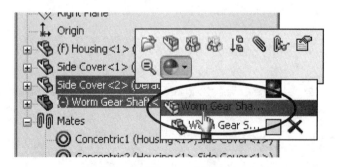

We can change a part's appearance in the assembly, or in the part file. Picture it like this: We can paint the part before we assemble it, or we paint it after we assemble it. In the second case, the part color changes only in the assembly. For this example we want to change the color at the assembly level, which is the option indicated above.

300. - We are now ready to change the part's color. Simply select the desired color from the color selection box to change it. Click OK to finish.

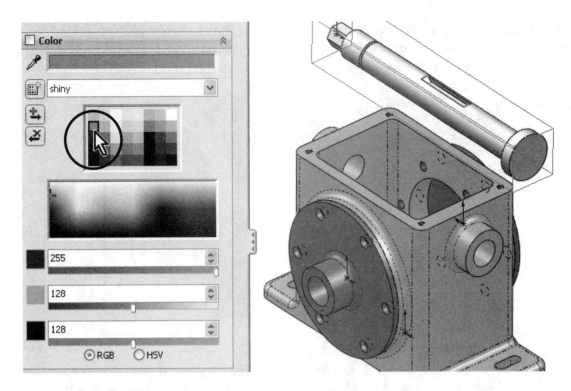

301. - Change the color of both covers to your liking and continue adding mates. (It does look better in color!)

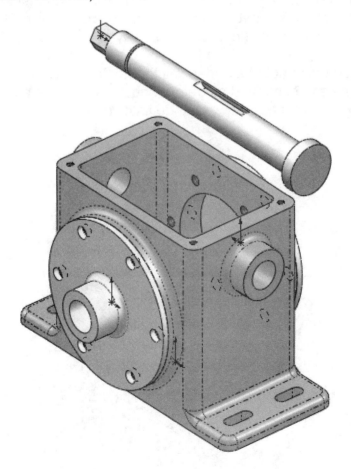

302. - Since we covered **Concentric** and **Coincident** mates and practiced them with the side covers, rotate the shaft approximately as the first image and add a Concentric and a Coincident mates using the faces indicated.

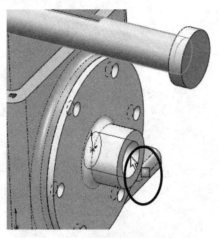
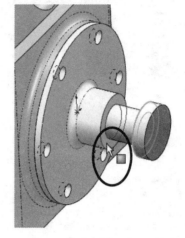

Concentric mate   Coincident mate

Notice that the Shaft is not yet fully defined, it still has a (-) sign before its name in the Feature Manager. In this case the only DOF left unconstrained is to rotate about the axis of the shaft, and that is exactly what we want, the shaft is supposed to rotate. If we click and drag the shaft, we'll see the shaft rotate (look at the keyway).

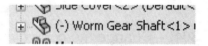

303. - The next component to be added will be the "Worm Gear". Insert it with the "Insert Component" command as before and place it in the assembly as shown. And now that you know how, change its color too.

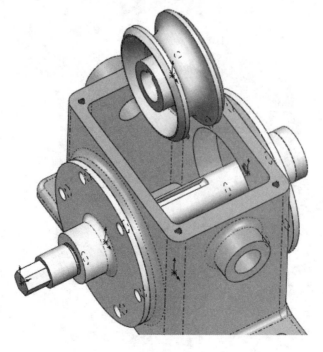

304. - To mate the "Worm Gear" in place, select the "**Mate**" icon and add a Concentric mate with the "Worm Gear Shaft" using the cylindrical inside face of the "Worm Gear" and the outside face of the shaft. Move the "Worm Gear" by dragging it just outside of the "Housing".

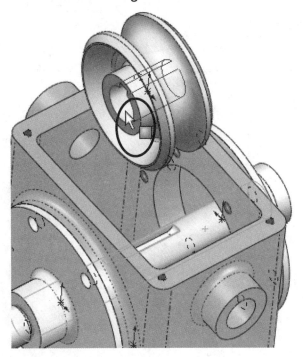

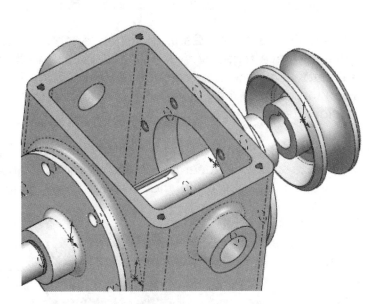

305. - Now we need to center the "Worm Gear" with the "Housing". To do this we can use the "Worm Gear's" Front plane and the Front plane of the "Housing" or the assembly, which are conveniently located in the center of the Housing. (Remember we made the "Housing" symmetrical about the origin? ☺)

Select the Mate command (if it's not open already) and add a **Coincident** mate selecting the Front Plane of the "Housing" (or the assembly) and the Front Plane of the "Worm Gear" from the fly-out Feature Manager. Expand the part's feature tree to select their planes.

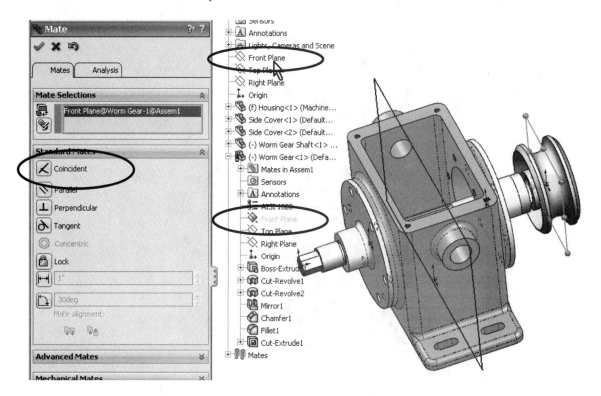

306. - After adding this mate the "Worm Gear" is still under defined, and we want the shaft and the gear to move together. To accomplish this we'll have to add either a **Coincident** mate using the corresponding planes from the parts, or a **Parallel** mate between the faces of the keyways. We'll do the Parallel mate here and let the reader explore the other option.

307. - As luck would have it, the "Housing" is obstructing the view to the keyways, and we'll have to hide it. Cancel the "**Mate**" command and select the "Housing" from the Feature Manger or in the graphics area, and select the "**Hide/Show**" icon from the pop up toolbar. The part will be hidden allowing us to see the rest of the components. We'll show it again after we are done mating the parts.

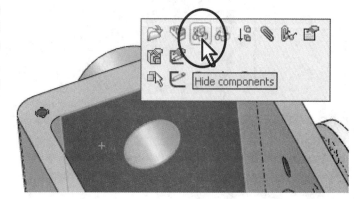

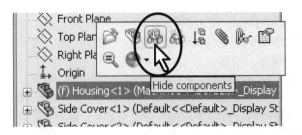

308. - As soon as we hide the part, it disappears from the screen. We can see inside without obstructions. The "Housing" icon in the Feature Manager changes to white; this is how we know that the part is hidden.

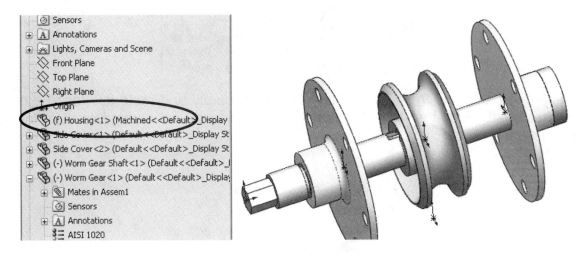

309. - Select the "Mate" icon again and select the flat faces of the shaft and gear keyways. In this case we'll use a "Parallel" mate. The **Parallel mate** is a good option, as it can help us absorb small dimensional differences that may exist between the parts. This is a more "forgiving" option. Click OK to apply the mate when finished.

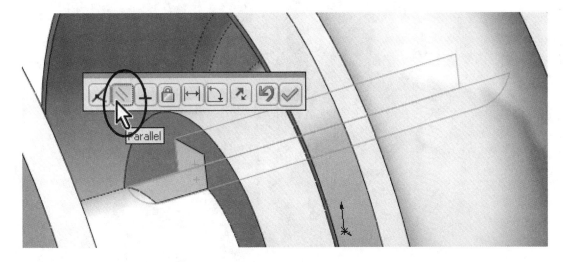

After adding the mate click and drag the gear or the shaft, we'll see both moving at the same time, as if they had the keyway inserted.

Extra credit: Make a keyway and add it to the assembly, mating the keyway to the shaft, and the Worm Gear to the keyway.

310. - To show the Housing again, we'll reverse the process that we used to hide it. Select the "Housing" from the Feature Manager and select the "Hide/Show Component" icon from the pop up toolbar. This command is really a toggle switch. If a component is hidden, you'll show it, if the component is visible, you'll hide it.

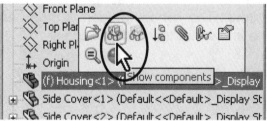

311. - Now let's add the "Offset Shaft" to the assembly and mate it. Insert the "Offset Shaft" using the "**Insert Component**" command and change its color for visibility.

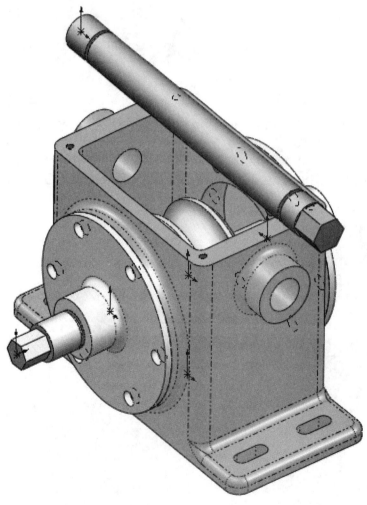

312. - Add mates by selecting one of the model faces to be mated, and from the pop up toolbar select the "**Mate**" icon. When the "Mate" command is displayed, this model face is already selected, now we only need to select the other face and add a Concentric mate.

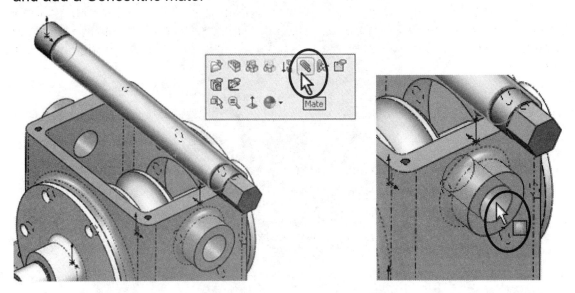

313. - Now we need a Coincident mate to prevent the shaft from moving along its axis. We'll use the groove in the shaft for this mate. In this case we can use either the flat face or the edge of the groove; selecting the edge may be easier than selecting the face. We can use the "Magnifying Glass" tool to zoom in and make our selections.

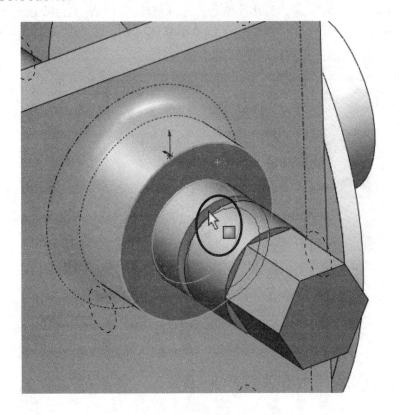

314. - Offset shaft assembled.

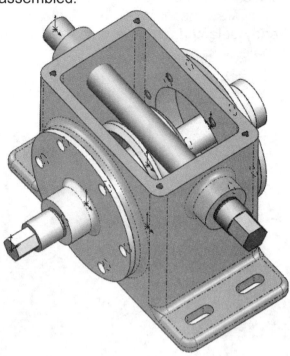

315. - The last component we're adding is the "Top Cover". Add it and if you wish, change its color. First add a **Concentric** mate to align one of the holes of the cover to a hole of the housing. For this mate you can use either faces or circular edges; it will work the same. An interesting detail to notice is that when we add a Concentric mate, it removes four DOF from the component, two translations and two rotations (if mated to a locked or fully defined reference).

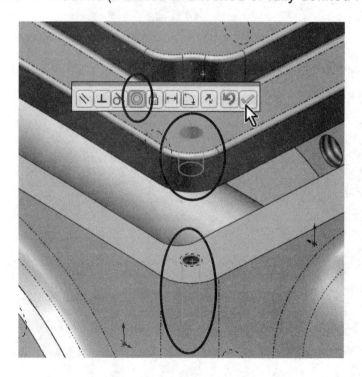

316. - Check the remaining two DOF by dragging the "Top Cover" with the left mouse button; it will rotate and move up and down. By adding a Coincident mate between the bottom face of the "Top Cover" and the top face of the "Housing" we'll remove one more DOF.

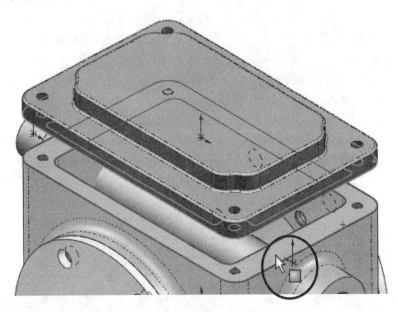

317. - If we now click and drag the "Top Cover" it will turn; now we have only one DOF left. To finish constraining the cover we'll add a **Parallel** mate between two faces of the "Top Cover" and "Housing". As we explained earlier, the reason for the Parallel mate is that sometimes components don't match exactly, and if we add a Coincident mate, we may be forcing a condition that cannot be met over defining the assembly getting an error message. The parallel mate can be added between faces, planes and edges. In this particular case we can use either Coincident or Parallel mate, since the parts were designed to match exactly. However, in real life they may not; that's why we chose a Parallel mate.

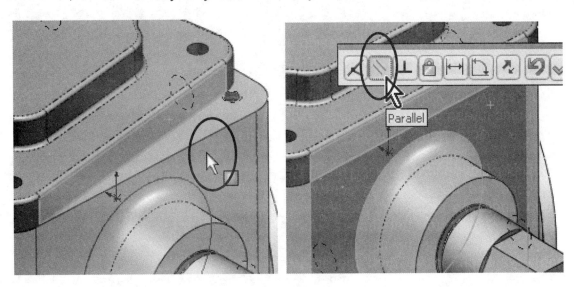

Your assembly should now look like this.

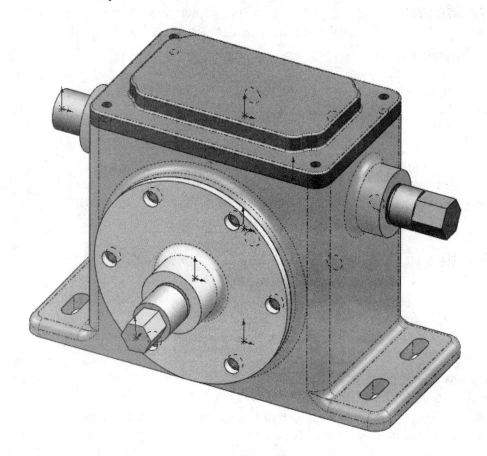

SmartMates

SmartMates are a quick and easy way to add certain mate types between components simply by dragging parts or assemblies using flat, cylindrical or conical faces, circular or linear edges, vertices or temporary axes. The basic idea is to hold down the "Alt" key while we drag the face, edge or vertex of the component to be mated *onto* the face, edge or vertex of the other component.

We'll re-create the entire assembly up to this point using only SmartMates, now that the concept of mates has been explained and we have a better idea of the general process. Here are the types of SmartMates available and their corresponding feedback icon.

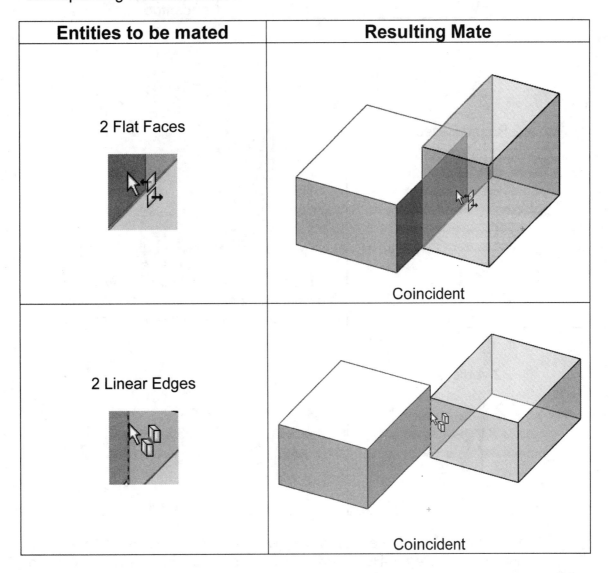

Entities to be mated	Resulting Mate
2 Flat Faces	Coincident
2 Linear Edges	Coincident

Entities to be mated	Resulting Mate
2 Cylindrical / Conical Faces	Concentric
2 Round Edges	**Peg-in-Hole** (Adds 1 Concentric mate between the cylindrical/conical faces and 1 Coincident Mate between the flat faces)
2 vertices	Coincident

 When using SmartMates to mate two flanged faces SolidWorks will consider it as a Peg-in-Hole, adding two concentric mates between the two pairs of cylindrical faces and one coincident mate between the two flat faces.

318. - Let's recreate the entire assembly now using SmartMates. Make a new assembly, and add the Housing as we did in the previous assembly. Go to the menu "New, Assembly", browse for the Housing and insert it in the origin. If you cannot see the origin, turn it on in the menu "View, Origins".

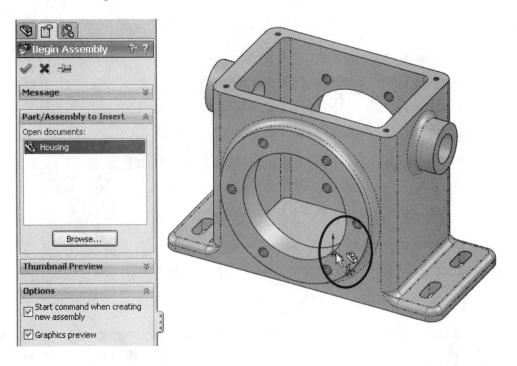

319. - Now bring in the Side Cover with the "Insert Component" command and put it next to the Housing. This is exactly as we did before up to this point.

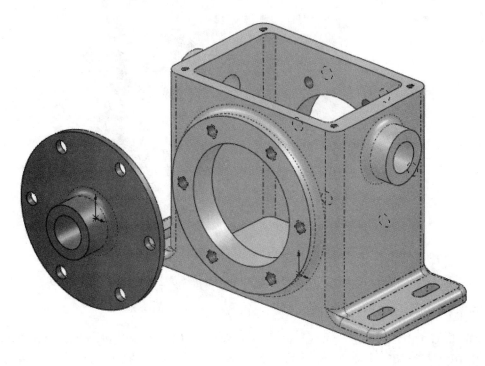

320. - Now we start using the SmartMates functionality. Press and hold the "Alt" key in the keyboard, and while holding it down left-click and drag the Side Cover from the edge indicated. Notice that as soon as we start moving it, the Mate icon appears next to the mouse pointer, and the Side Cover becomes transparent.

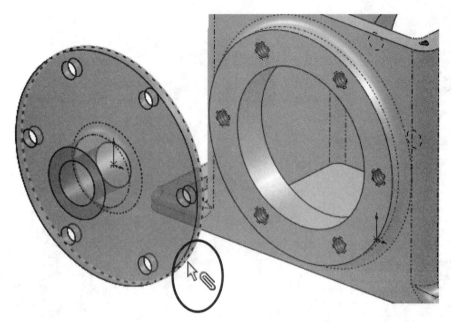

321. - Keep dragging it until we touch the flat face or the round edge in the front face of the Housing; at this time the Side Cover will 'snap' into place and will give us the "**Peg-in-Hole**" icon. Release the mouse button to finish. The Side Cover is now fully defined to the Housing with a single step.

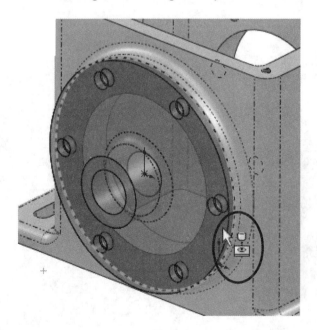

 Expanding the "Mates" folder in the Feature Manager we can see that two concentric mates and one Coincident mate have been applied. Select the mates to see the faces that were mated in each one.

322. - Even as we can add the second Side Cover easily with SmartMates, we'll show how to add it by using the "**Mirror Components**" command. Select the menu "Insert, Mirror Components".

323. - Similar to the Mirror Feature command, we need a plane of flat face to be used as a mirror plane. In this case we can select the assembly's Front plane as the Mirror Plane (since it is located in the middle of the assembly), and in the "Components to Mirror" selection box select the Side Cover. After making our selections click in the "Next" blue arrow to go to the next step in the Mirror Component command.

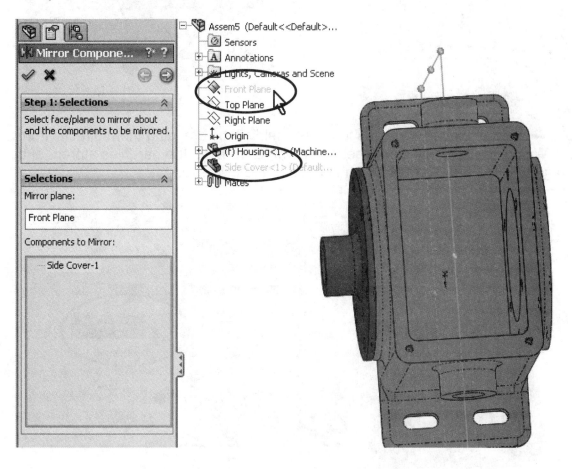

324. - In this step we get a preview of the mirrored part and we can optionally change its orientation and/or create an "opposite hand" version of it. In this case the default orientation is correct and we don't need an opposite hand version. Click OK to finish. The new side Cover is added and a "Mirror Component" feature is added to the Feature Manager.

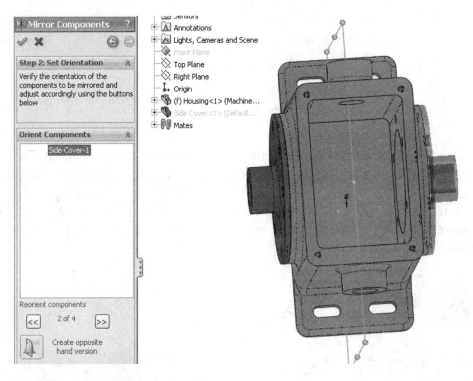

325. - Another way to add parts to an assembly is by dragging them directly from the Windows Explorer. Open the folder where we have the part files stored, and drag-and-drop the Worm Gear Shaft directly into the assembly.

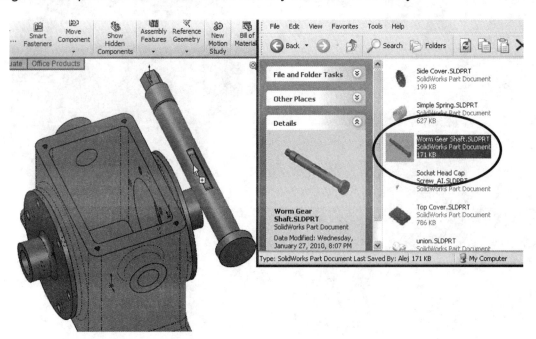

326. - After adding the Worm Gear Shaft to the assembly, rotate the shaft to a position that will allow us to view both of the edges that we want to mate. SmartMates has a limitation in the sense that we need to be able to see both of the mate references, in this case the edge indicated in the left image and the Side Cover's center hole. Hold down the "Alt" key, click and drag the shaft's edge to the Side Cover to add a "Peg-in-Hole" SmartMate.

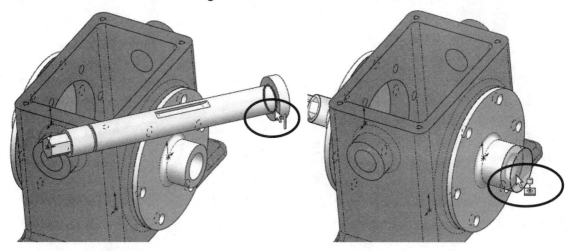

327. - Another way to add a component to an assembly is by opening it in SolidWorks, arranging the part and assembly windows side by side (menu "Window, Tile Vertically"), and dragging the part into the assembly. One advantage of this method is that dragging the part using the face that we want to mate into the assembly activates the SmartMates function automatically.

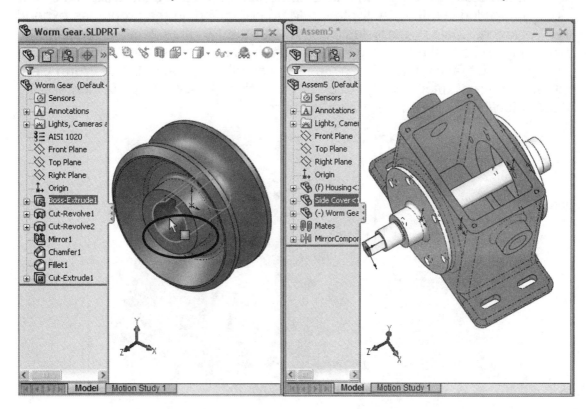

328. - As we drag the part into the assembly window move the Worm Gear to the shaft to add a concentric mate automatically (look for the Concentric mate icon).

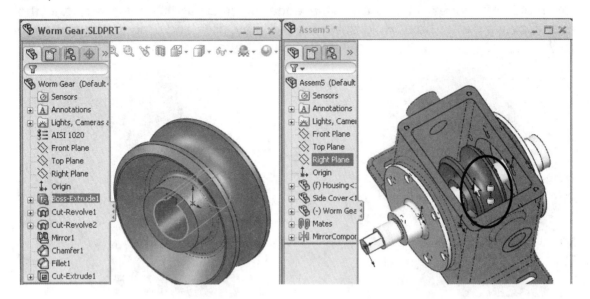

 We can drag parts into assemblies using faces, edges or vertices, just as we would for SmartMates. The only difference is that in this case we don't have to press the "Alt" key.

329. - The next step is to center the Worm Gear. In the previous assembly we added a mate using the part and assembly's planes. In this case we'll use a new mate type called "**Width**". Activate the Mate command; in the properties expand the "Advanced Mates" and select "Width". In the "Width selections:" select the two inside faces of the Housing, click in the "Tab selections:" box and select the two outside flat faces of the Worm Gear (the small round faces). The Worm Gear will be centered automatically. Click OK to finish. The Width mate centers the "Tab" selections between the "Width" selections.

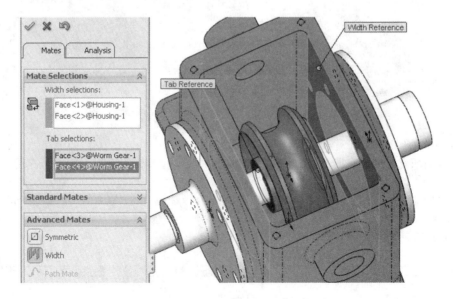

330. - Add the Offset Shaft to the assembly and mate it using SmartMates.

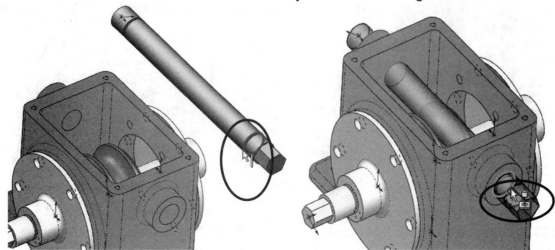

331. - Add the Top Cover and flip it as shown in the next picture. Note that the Top Cover has been turned upside down. One reason for it is to be able to see both of the edges to mate, the other to show how to flip a SmartMate.

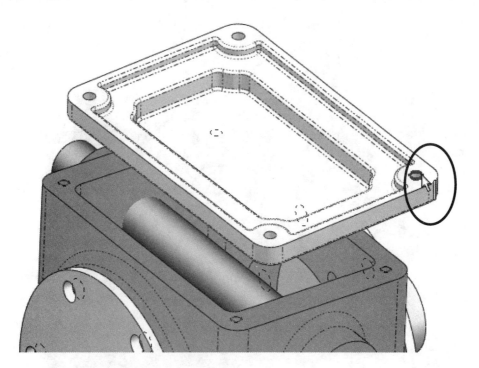

332. - Select the edge indicated, and add a Peg-in-Hole SmartMate. Note that the cover may be upside down in the preview. **BEFORE** RELEASING THE MOUSE BUTTON, release the "Alt" key, and press the "TAB" key once, notice how the preview changes flipping the cover. Pressing "TAB" again will flip the mate again. Make sure you have the correct orientation for the Top Cover and release the mouse button to add the SmartMate.

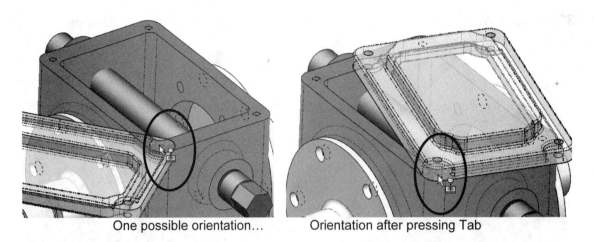

One possible orientation…   Orientation after pressing Tab

333. - Now add a SmartMate to finish locating the Top Cover. Drag the edge of the Top Cover and SmartMate it to the face (or the edge) of the Housing to make them coincident. This will fully define the Top Cover.

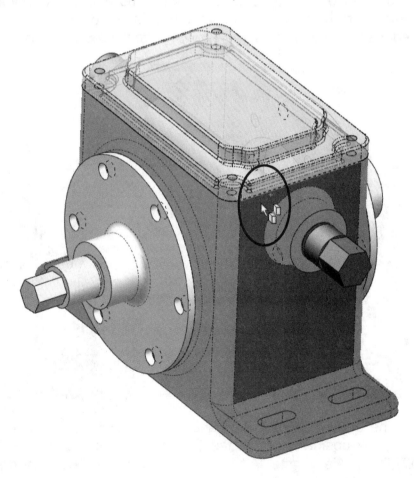

The reader can see why it's a good idea to learn how to use SmartMates, as it helps to speed up the assembly process and making the process dramatically easier.

Fasteners

The commercial version of SolidWorks Office Professional and the educational edition available to schools include a library of hardware called "**SolidWorks Toolbox**" that includes nuts, bolts, screws, pins, washers, bearings, bushings, structural steel, gears, etc. in metric and inch standards. The Toolbox is an accessory that has to be loaded through the menu "**Tools, Add-ins**". For Toolbox to work correctly we have to load "SolidWorks Toolbox" and "SolidWorks Toolbox Browser".

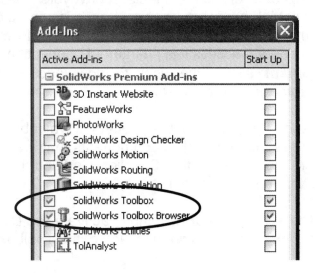

With the SolidWorks Toolbox we can add hardware to our assemblies by simply dragging and dropping components, and SolidWorks will automatically add the necessary mates saving us time. We can also add our own hardware to it, making it more versatile.

334. - To access the SolidWorks Toolbox we use the **Design Library**, which is located in one of the tabs in the Task Pane. If the Task Pane is not visible, go to the menu "**View, Toolbars, Task Pane**" to activate it.

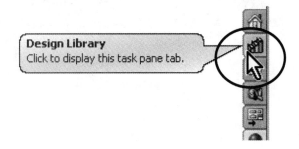

 The SolidWorks Toolbox was not included in the Student Design Kit as of the writing of this book.

Clicking on the Design Library icon opens a fly-out pane that reveals the library. The Design Library contains four main areas:

- **Design Library**, which includes built-in and user defined libraries of annotations, features and parts that can be dragged and dropped into parts, drawings and assemblies.
- **Toolbox**, which we just described.
- **3D Content Central**, an Internet based library of user uploaded and manufacturer certified components, including nuts, bolts and screws, pneumatics, mold and die components, conveyors, bearings, electronic components, industrial hardware, power transmission, piping, automation components, furniture, human models, etc., all available for drag and drop use. All that is needed to access it is an Internet connection.
- **SolidWorks Content** allows the user to download weldments libraries, piping, blocks, structural members, etc.

As we can see, the Design Library offers a valuable resource for the designer, helping us save time modeling components that are usually purchased or standard, and in the case of the Supplier Certified library, components accurately modeled for use in our designs.

335. - To start adding screws to our assembly, activate the Design Library from the Task Pane, and click on the (+) sign to the left of the Toolbox to expand it.

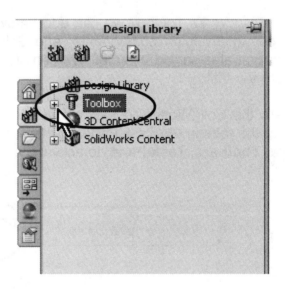

336. - After expanding the Toolbox we can see the many options available. Depending on how Toolbox is configured, some standards may not be available. In this exercise we'll use "ANSI Inch", "Bolts and Screws" and select "Socket Head Screws".

When we click the "**Socket Head Screws**" folder we see in the lower half of the Design Library pane the available styles, including Button Head, Socket Head, Countersunk and Shoulder Screws. For our assembly we'll use Socket Head Cap Screws (SHCS).

337. - First we want to add the #6-32 screws to hold the Top Cover. From the bottom pane of the Toolbox, click-and-drag the Socket Head Cap Screw into one of the holes in the Top Cover. You will see a transparent preview of the screw, and when we get close to the edge of a hole, SolidWorks will automatically snap the screw in place using SmartMates. When we get the preview of the screw assembled where we want it, release the left mouse button. If we release the mouse button before, the screw will still be created, but it will not be automatically assembled or add any mates.

 Do not worry if the preview in your screen is big. When we drop the screw in place it will size to match the hole.

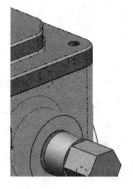 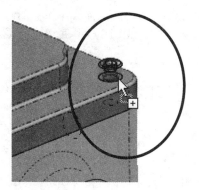

Notice that the Design Library hides away as soon as we drag the screw in the assembly; this is an automatic feature, unless the user pressed the "stay visible" thumb tack on the upper right of the Task Pane.

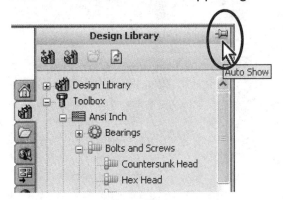

Auto-hide On   **Always visible**

338. - As soon as we drop the screw in the hole, we are presented with a dialog box asking us to select the screw parameters and a pop up menu where we can select the screw size. In this case, we need a #6-32 screw, 0.5" long with Hex Drive and Schematic Thread Display.

A word on Thread Display: The schematic thread display will add a revolved cut to the screw for visual effect. Helical threads can be added as a feature, but it's generally considered a waste of computer resources that doesn't really add value to a design in most instances. Helical threads are a resource intensive feature that is best left for times when the helical thread itself is a part of the design and not just for cosmetic reasons. The revolved cut gives a good appearance for most practical purposes and is a simple enough feature that doesn't noticeably affect the PC's performance.

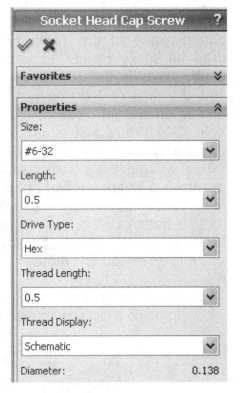

Screws are stored in a master file that includes the different configurations of each screw type in the default Toolbox data folder. If the assembly files are copied to a different computer the screws used will be created there. If the other computer does not have Toolbox, the user can copy all the files using the menu option "File, Pack and go" to copy all the files used in the assembly, including Toolbox components and drawings.

339. - After selecting OK from the screw size options box, the screw is created with the selected parameters and mated in the hole where it was dropped. At this point we are ready to add more screws of the same size if needed. In our example, we'll click in the other 3 holes of the Top Cover to add the rest of the screws. Notice the graphic preview of the screw as we move the mouse and the screw snaps as we get close to a hole adding a "Peg-in-Hole" SmartMate. Click "Cancel" to finish adding screws.

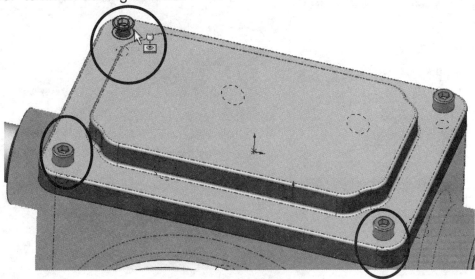

340. - We will now add the ¼-20 Socket Head Cap Screw to the side covers. Drag and drop the Socket Head Cap Screw to one hole of the Side Cover, but be careful to "drop" the screw in the correct location. If you look closely, there are two different edges where you can insert the screw: one is in the Side Cover, and the other is in the Housing. The screw will be mated to the hole you "drop" the screw in. Use the snap preview to help you find the correct one.

341. - After we drop the screw, the screw properties box is displayed. Select the ¼-20, 0.5" long screw with hex drive and schematic thread display. Click OK to make the screw and mate it in the hole where we dropped it. Insert six screws on one Side Cover. As a challenge, add the other six screws using the Mirror Components command.

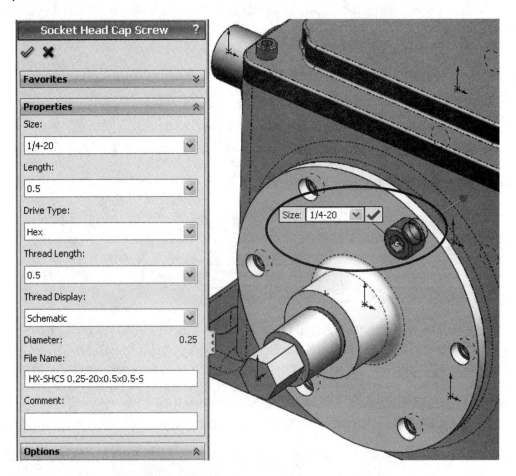

342. - For display purposes, we can turn off the component origins and cosmetic threads. Go to the menu "**View**", and turn off the "**Origins**" option. For the cosmetic threads, Right-Mouse-Click in the "**Annotations**" folder in the Feature Manager, select "Details" and turn off the "Cosmetic Threads" option. Make sure to select the "Use assembly setting for all components" option.

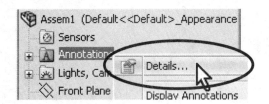

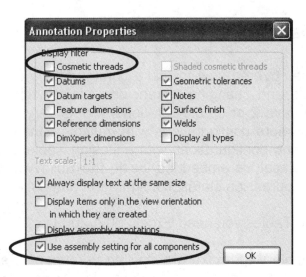

Save the assembly as "Gear Box".

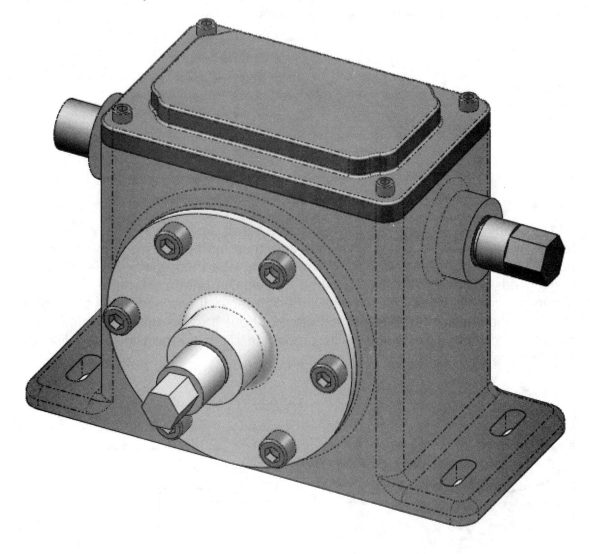

Configurations using Design Tables

Previously we covered how to make configurations of parts by manually suppressing features and changing dimensions. While this approach may be adequate when we have 2 or 3 configurations, it gets very complicated to keep track of 5, 10, 20 or more configurations with multiple dimensions and features. This is when controlling configurations with a Design Table is a good idea. The Design Table is an Excel file embedded inside the SolidWorks file that controls dimension values, suppression states, configuration names, etc.

 To use Design Tables you need Excel 2002, 2003 or 2007 installed.

343. - For this example we'll make a simplified version of a screw, with only a handful of sizes. Before we make the design table, we need to have a model to configure. Please make the following sketch in the Front plane. Remember to add the centerline for the diameter dimensions.

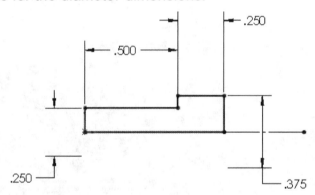

344. - SolidWorks assigns an internal name to all dimensions with the general format *name@feature_name.* We can see their name by resting the mouse pointer on top of a dimension, or if we want to see them always (Useful when making design tables), go to the menu "**View, Dimension Names**". When we configure dimensions, especially with a design table, it is a good idea to rename them to something that tells us something about it. To rename a dimension select it and type a new name in its properties. Turn on the Dimension Names and rename them as shown.

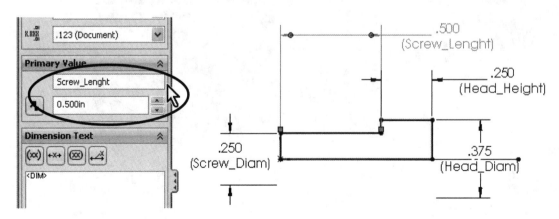

345. - Make a revolved boss. After making a feature the sketch dimensions are hidden. To make model dimensions always visible, even when editing the part Right-Mouse-Click in the "Annotations" folder in the Feature Manager, and activate "Show Feature Dimensions".

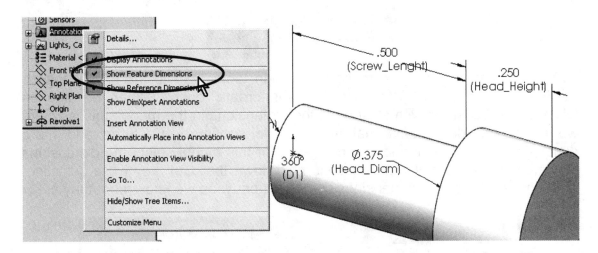

346. - Add a hexagonal cut in the head of the screw, make it 0.125" deep.

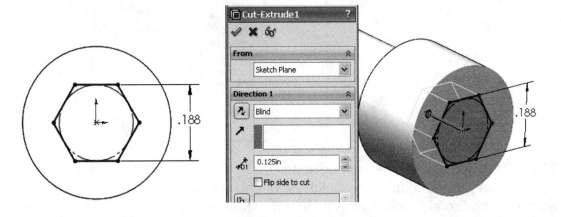

347. - Change the cut's dimension names to "Hex_Drive" and "Hex_Depth". Note that dimensions added in the sketch are shown in black and the ones added by the feature are blue.

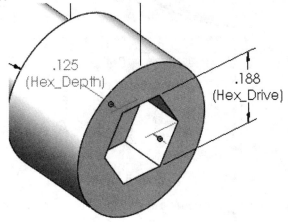

348. - Add a 0.015" x 45 deg. chamfer to the head and tip of the screw.

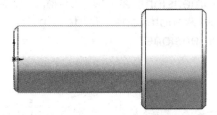

349. - Now we are ready to make the design table. Go to the menu "**Insert, Tables, Design Table**". We'll use the "Auto-create" option. When we click on OK, we are presented with a list of all the dimensions in the model; this is where we select the dimensions that we want to configure. Features and more dimensions can be added later if needed. For this model select the dimensions indicated. To make multiple selections hold down the "Ctrl" key while selecting. This is where renaming dimensions comes in handy, as we know what they are.

350. - After selecting the dimensions click OK. An Excel spreadsheet is automatically embedded in the SolidWorks part and we are ready to edit it. Since we are using Excel *inside* SolidWorks, our menus and toolbars change to Excel.

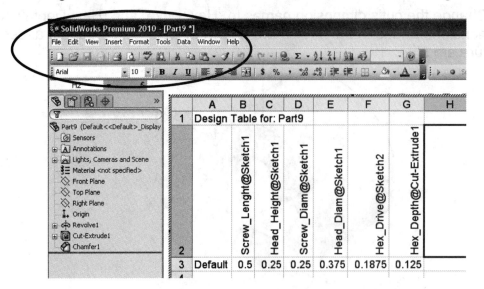

351. - The standard behavior of embedded documents in MS Windows is to add a thin border around the embedded document. If needed, we can move the Excel file by dragging this border, or resize it from the corners. Be aware that they are very small, be careful not to click outside, or you'll get back to SolidWorks.

IMPORTANT: If you accidentally click outside the Excel spreadsheet, this is what will happen: You may be told that that a configuration was created or not, depending if you modified the design table. To go back to editing the design table, go to the Configuration Manager, expand the "Tables" folder, make a right click in the "Design Table" icon and select "Edit Table". This will get you back.

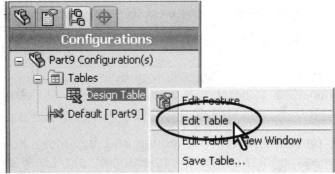

352. - Resize the table and columns if needed. As we always have at least one configuration, "Default" is listed in our design table with the corresponding values listed under each parameter. Our design table that now looks like this:

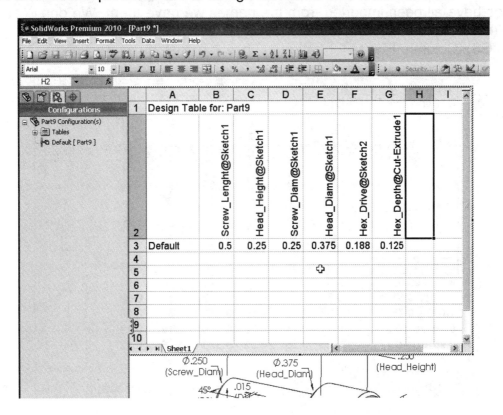

353. - The Design Table holds the information of the parameters that are configured, the configuration names and values for each parameter in every configuration. The first Row holds information about the part. We can type anything we want in this row, as the configuration's data start in the second row. Configurable parameters are listed in the second row starting at the second column, and configuration names are listed in the first column starting in the third row. Here is a list of the most commonly configured parameters in a part:

Parameter	Format in Design Table	Possible Values
Dimensions (To control dimension value)	*name@feature_name*	Any decimal value
Features (To control if feature is suppressed or not)	*$STATE@feature_name*	S or *Suppressed* U or *Unsuppressed*
Custom properties (To add custom properties to a configuration)	*$prp@property* property is the name of the custom property to add	Any text string

 Many other parameters can be configured including hole wizard sizes, description, mass, center of gravity, tolerances, etc. Look at the SolidWorks help for "Summary of Design Table Parameters" for more details.

354. - Now we'll fill the values for the new configurations to be added. Edit the table as show; note that "Default" configuration was renamed. The dimension names have all been imported from the names we gave them. We don't need to type the dimension names, just the Configuration names and their values (Cells A3 to G8). Column "F" was formatted to show dimensions as fractions.

	A	B	C	D	E	F	G	H	I
1	Design Table for: Part9								
2		Screw_Lenght@Sketch1	Head_Height@Sketch1	Screw_Diam@Sketch1	Head_Diam@Sketch1	Hex_Drive@Sketch2	Hex_Depth@Cut-Extrude1		
3	6-32 x 0.5	0.5	0.138	0.138	0.226	7/64	0.064		
4	6-32 x 0.75	0.75	0.138	0.138	0.226	7/64	0.064		
5	10-32 x 0.5	0.5	0.19	0.19	0.312	5/32	0.09		
6	10-32 x 0.75	0.75	0.19	0.19	0.312	5/32	0.09		
7	0.25 -20 x 0.5	0.5	0.25	0.25	0.375	3/16	0.12	⊕	
8	0.25 -20 x 0.75	0.75	0.25	0.25	0.375	3/16	0.12		
9									
10									

Sheet1

355. - Now that the design table is complete, click anywhere inside the graphics area of SolidWorks to exit the design table (and Excel). We will be told that the configurations have been created. Click OK to continue.

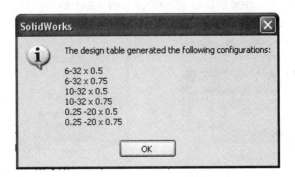

356. - Looking at the Configuration Manager, we can see the added configurations. Note that the configurations created with the Design Table have an Excel icon next to them. The configured dimensions are shown with a magenta color (Default setting). Switch to other configuration to see the changes.

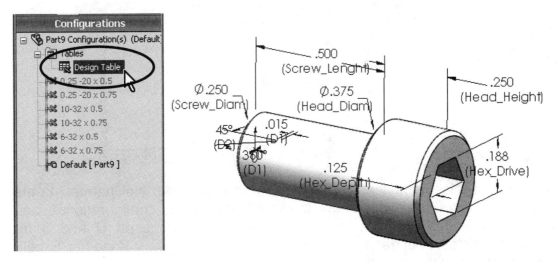

 We can delete the "Default" configuration if we want to, but we have to change to a different configuration, we can't delete the active configuration.

357. - To add features edit the design table. Right mouse click in the "Design Table" and select "Edit Table". If asked to add a configuration or parameters click on "Cancel".

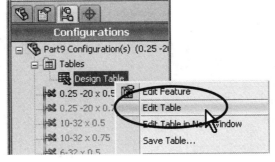

358. - When the table is presented, the next available cell for adding a new parameter is pre-selected (in our case "H2"). To add the chamfer as a configurable parameter, we can type *$STATE@Chamfer1* directly in the cell, or select the Feature Manager's tab to view the features tree, and double click in the "Chamfer1" Feature.

 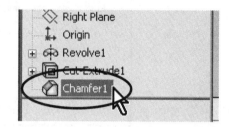

This will add the correct nomenclature in the table and its current suppression state. In this case Unsuppressed. If we leave the value empty SolidWorks will assume Unsuppressed. For short we can type S for Suppressed or U for Unsuppressed.

	A	B	C	D	E	F	G	H
1	Design Table for: Part9							
2		Screw_Lenght@Sketch1	Head_Height@Sketch1	Screw_Diam@Sketch1	Head_Diam@Sketch1	Hex_Drive@Sketch2	Hex_Depth@Cut-Extrude1	$STATE@Chamfer1
3	6-32 x 0.5	0.5	0.138	0.138	0.226	7/64	0.064	UNSUPPRESSED

359. - Copy the last configuration to the next row and add "NoChamfer" to the name (SolidWorks does not allow duplicate names for configurations), and suppress "Chamfer1" just as an exercise. For short we used **U** and **S** for suppression states. Feel free to format the Design Table to your liking (it's Excel), as long as you don't change the layout.

	A	B	C	D	E	F	G	H
1	Design Table for: Part9							
2		Screw_Lenght@Sketch1	Head_Height@Sketch1	Screw_Diam@Sketch1	Head_Diam@Sketch1	Hex_Drive@Sketch2	Hex_Depth@Cut-Extrude1	$STATE@Chamfer1
3	6-32 x 0.5	0.5	0.138	0.138	0.226	7/64	0.064	U
4	6-32 x 0.75	0.75	0.138	0.138	0.226	7/64	0.064	U
5	10-32 x 0.5	0.5	0.19	0.19	0.312	5/32	0.09	U
6	10-32 x 0.75	0.75	0.19	0.19	0.312	5/32	0.09	U
7	0.25 -20 x 0.5	0.5	0.25	0.25	0.375	3/16	0.12	U
8	0.25 -20 x 0.75	0.75	0.25	0.25	0.375	3/16	0.12	U
9	0.25 -20 x 0.75(NoChamfer)	0.75	0.25	0.25	0.375	3/16	0.12	S

360. - After finishing the changes in the design table return to SolidWorks just as we did before. We'll get a message letting us know that a new configuration was created. Click OK to finish.

361. - Hide the model dimensions and save the part as "Screw Design Table". The finished configurations look like this:

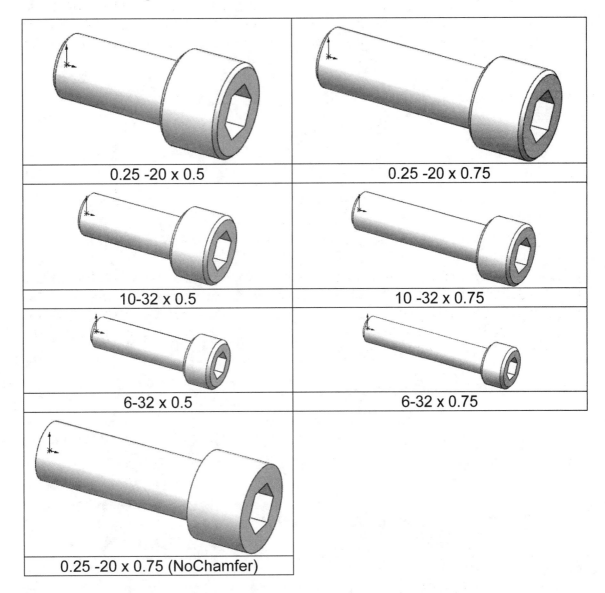

0.25 -20 x 0.5	0.25 -20 x 0.75
10-32 x 0.5	10 -32 x 0.75
6-32 x 0.5	6-32 x 0.75
0.25 -20 x 0.75 (NoChamfer)	

Notes:

Interference Detection

362. - Back to the Gear Box assembly. One tool that will help us find problems in our assembly is the "**Interference Detection**" command. We can find it in the "Evaluate" toolbar, or in the menu "**Tools, Interference Detection**". We have a designed interference in the assembly to show the user how to use this tool. Click on "Interference Detection" and click on "Calculate". The entire assembly should be selected by default.

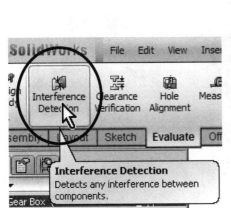

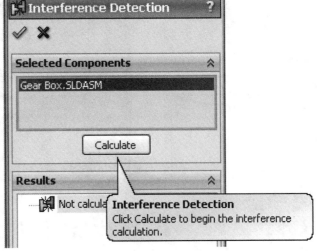

363. - After calculating the interferences we see many interferences listed in the "Results" window. Most of these are fasteners, because the holes they fit in are smaller than the screw threads. Since we know that, we are going to use an option called "Create fasteners folder". When we activate it, all the fastener related interferences will be grouped together, making it easier to identify problem areas that are not fasteners. Turn this option on. You may also find it useful to turn on the "Hidden" option in the "Non-interfering Components" box to temporarily hide all the components that are not related to the interference making the problem areas easier to find.

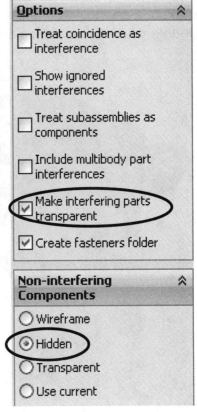

364. - Select an interference of the list to see the interfering volume in the screen. Now the designer knows where the problem is, and can take corrective action. This is a tool that will help us make better designs, but it will only show where the problems are, it's up to the designer to fix them.

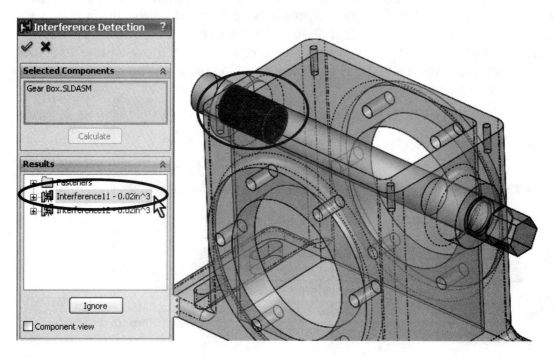

365. - In our case, we know the hole is smaller than the shaft, so we need to make the shaft smaller. To do this, exit the Interference Detection command, and double-click on the shaft's cylindrical surface to reveal the feature's dimensions.

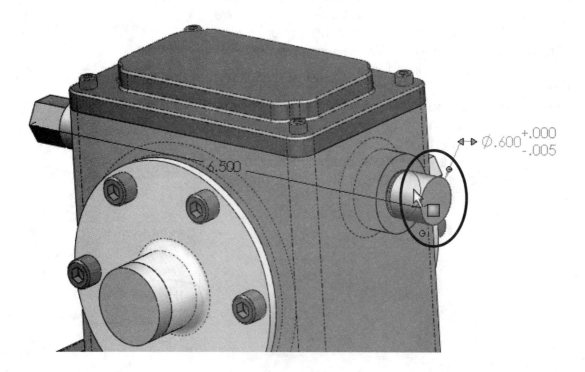

366. - Double-click in the diameter dimension and change its value to 0.575". Click in the "Rebuild" icon as shown, and then OK to complete the "Modify" command. Rebuilding the model will tell SolidWorks to update any models that were changed like the shaft.

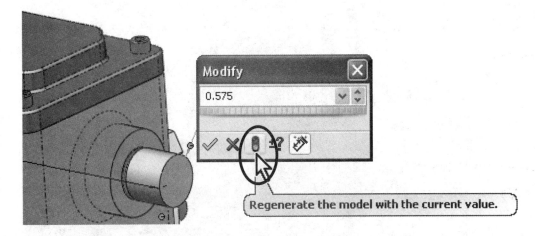

367. - Run the "Interference Detection" command again to confirm that we don't have any more problems.

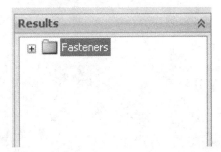

368. - Open the "Offset Shaft" drawing and notice the shaft's diameter dimension has been updated. Save and close the drawing file. When asked to save modified files click in "Save All" to save both the part and the drawing.

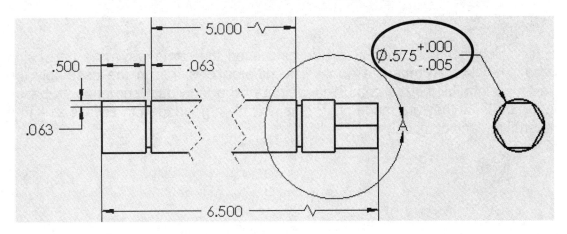

369. - Your Gear Box assembly is finished and should now look like this.

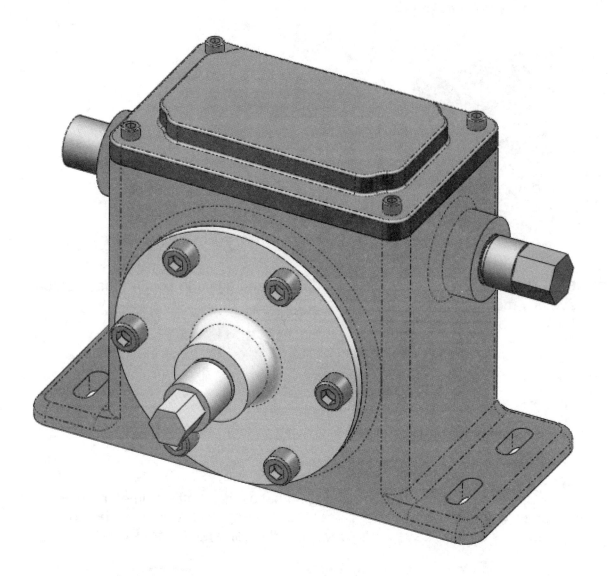

An assembly's weight can be calculated the same way as a part using the "Mass Properties" tool. The difference is that in the assembly the weight will be the combined weight of the individual components based on the material they are made of. This is why it's a good idea to always assign a material to each component.

Exploded View

370. - The last step in the assembly will be to make an exploded view for documentation purposes. Exploded views are used to show how the components will be assembled together. Select the "**Exploded View**" icon from the "Assembly" tab or from the menu "**Insert, Exploded View**".

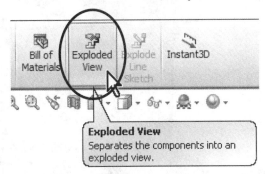

371. - In the Explode Property Manager we can see the "Settings" selection box is active and ready for us to select the component(s) that will be exploded. For this example we will leave the option "Auto-space components after drag" off.

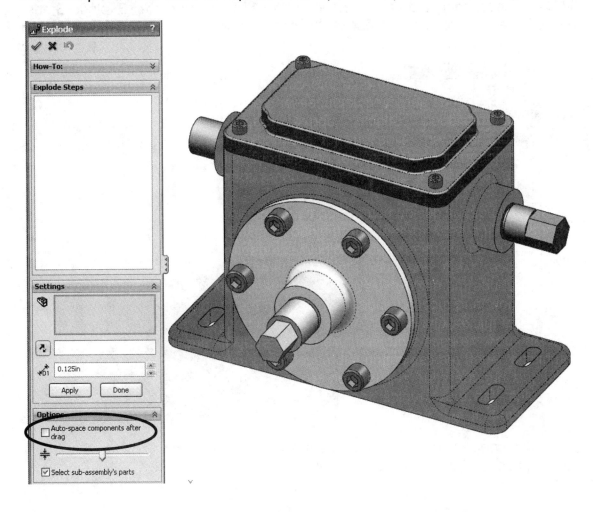

372. - Adding exploded view steps is very simple and straightforward. For the first explode step, select the four screws on the "Top Cover" and drag the green tip of the manipulator arrow upwards as far as you want the screws exploded. Note the ruler as we drag the screws up.

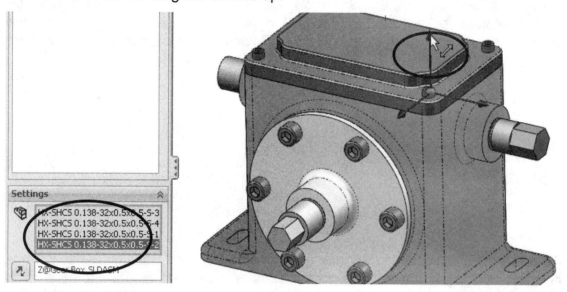

373. - After dragging the arrow, "Explode Step1" is added to the "Explode Steps" list, and the selection box is cleared. Do not click on OK yet, we are going to add more explode steps.

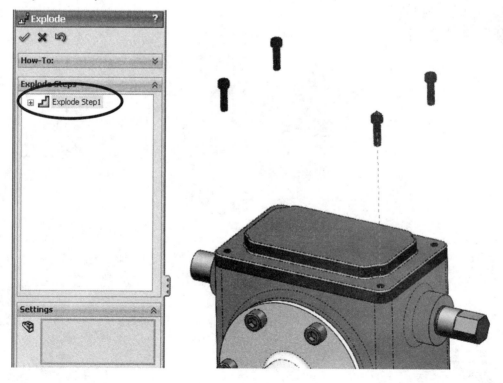

 If we need to modify an explode step select it from the "Explode Steps" list and drag the blue manipulator arrow to the new location.

374. - For the second explode step, select the "Top Cover" and drag the green manipulator arrow up about halfway between the screws and the "Housing".

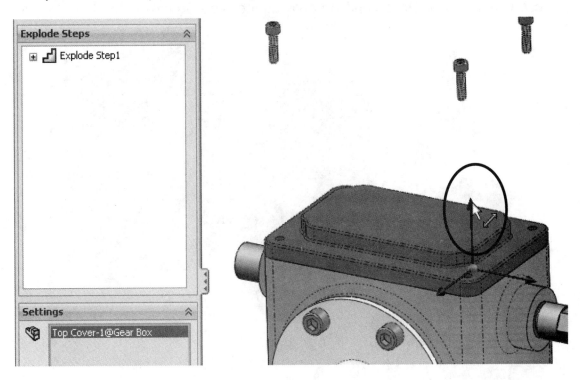

375. - For the third step, select the six screws on one of the "Side Covers" and drag them to the left using the blue manipulator arrow.

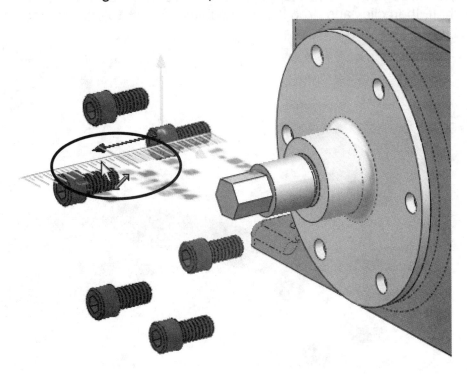

376. - The rest of the explode steps are done the same way, selecting the component(s) and dragging the tip of the manipulator arrow along the explode direction desired. The next step is to explode the Side Cover.

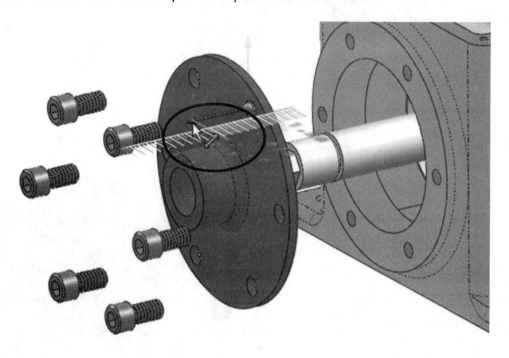

377. - Now explode the "Worm Gear Shaft" and the screws from the other side of the gear assembly in one step. Rotate the assembly and select the screws and the shaft, then click and drag the manipulator's arrow to explode them.

378. - Now explode the second "Side Cover".

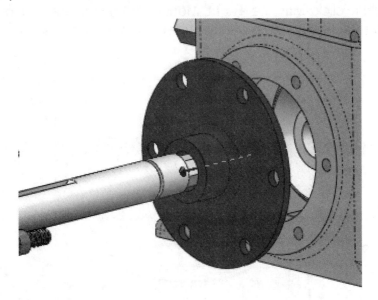

379. - Finally explode the "Worm Gear" and the "Offset Shaft" in two different steps to get the exploded view in the next image. Try to group parts that will be exploded the same distance and direction at the same time to reduce the number of steps required to document our design. When we finish adding explode steps click OK to finish the Exploded view command.

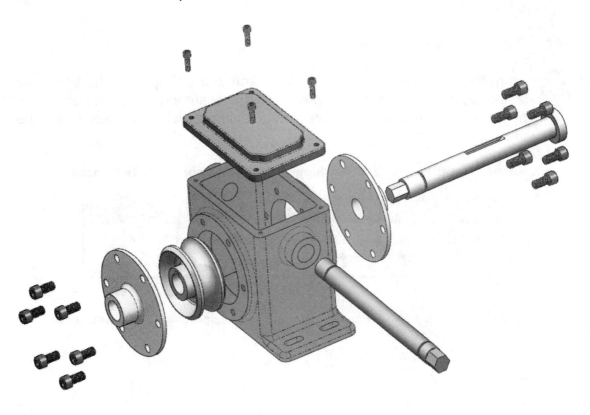

380. - After exploding the assembly, we may want to collapse the assembly. In order to do this, click with the Right Mouse Button at the top of the Feature Manager in the Assembly name and select "**Collapse**" from the pop up menu.

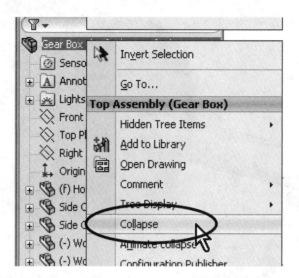

381. - To explode the assembly again, click with the Right Mouse Button at the top of the Feature Manager as we just did, and select "**Explode**". This option will only be available if the assembly has been exploded. In case we need to edit the exploded view steps again, select the "**Exploded View**" icon to bring back the explode Property Manager.

382. - Save the "Gear Box" assembly.

 From this same menu, we can select "**Animate collapse**" if the assembly is exploded, or "**Animate explode**" if the assembly is collapsed. This brings up the Animation Controller to animate the explosion using the animation controls.

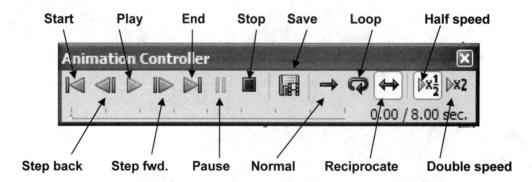

Assembly Drawing

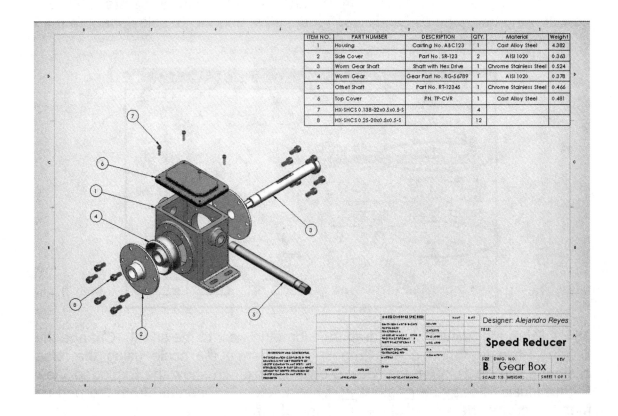

ITEM NO.	PART NUMBER	DESCRIPTION	QTY.	Material	Weight
1	Housing	Casting No. ABC123	1	Cast Alloy Steel	4.382
2	Side Cover	Part No. SR-123	2	AISI 1020	0.363
3	Worm Gear Shaft	Shaft with Hex Drive	1	Chrome Stainless Steel	0.524
4	Worm Gear	Gear Part No. RG-56789	1	AISI 1020	0.378
5	Offset Shaft	Part No. RT-12345	1	Chrome Stainless Steel	0.466
6	Top Cover	PN. TP-CVR	1	Cast Alloy Steel	0.481
7	HX-SHCS 0.138-32x0.5x0.5-S		4		
8	HX-SHCS 0.25-20x0.5x0.5-S		12		

Designer: *Alejandro Reyes*

TITLE:

Speed Reducer

SIZE DWG. NO.   REV

B Gear Box

SCALE 1:5 WEIGHT: SHEET 1 OF 1

383. - After making the assembly it is usually required to make an assembly drawing with a Bill of Materials for assembly instructions and documentation. In this case we will make a drawing with an exploded view, Bill of Materials (BOM) and identification balloons. To make the drawing, open the assembly "Gear Box". Select the "Make drawing from part/assembly" icon as before. Make a new drawing using the "B-Landscape" sheet size, and the "Display sheet format" checkbox activated.

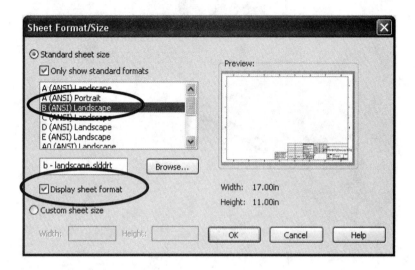

384. - From the "View Palette" drag the isometric view to the sheet. Change the Display Style to "Shaded with Edges" and change the scale using the "Use custom scale" to 1:2 from the "Scale" options box. To show the Exploded View Right Mouse Click in the isometric view and select "Properties" from the pop up menu. Activate the option "Show in Exploded State".

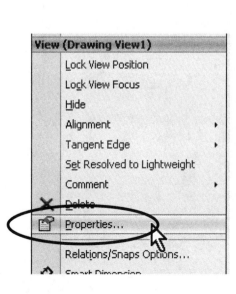

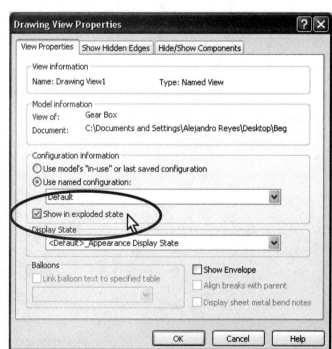

385. - Once we have the exploded Isometric view in the drawing, we need to add a **Bill of Materials** (BOM). Select the isometric view from the graphics area by clicking on it, and select the menu "**Insert, Tables, Bill of Materials**", or from the right mouse button menu (in the isometric view) "Tables, Bill of Materials".

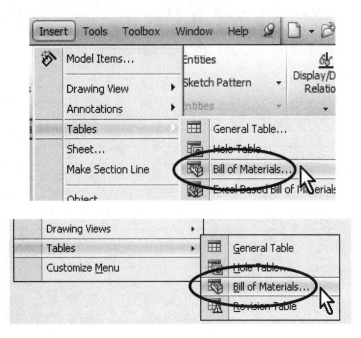

386. - Click OK in the Bill of Materials Property Manager to accept the default options.

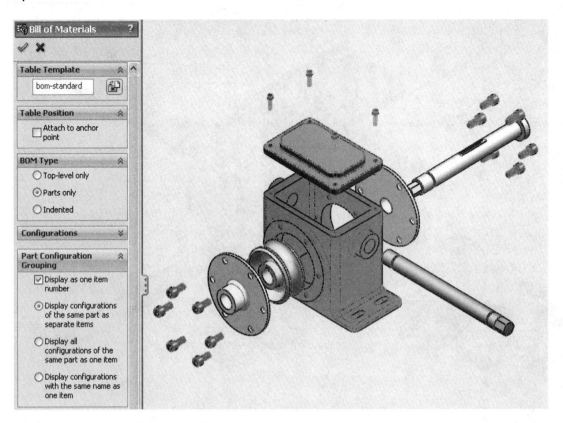

387. - After clicking on OK we can locate the Bill of Materials on the sheet. Move the mouse and locate the BOM in the top right part of the drawing. While locating the table notice the table will snap to the corner; be sure to allow some space for the identification balloons that will be added later. If needed, move the assembly view. The column's width and row's height can be adjusted by dragging the table lines as in Microsoft Excel. Since we added the "Description" custom property to the Worm Gear Shaft, Worm Gear and Offset Shaft, they are imported into the drawing.

ITEM NO.	PART NUMBER	DESCRIPTION	QTY.
1	Housing		1
2	Side Cover		2
3	Worm Gear Shaft	Shaft with Hex Drive	1
4	Worm Gear	Gear Part No. RG-56789	1
5	Offset Shaft	Part No. RT-12345	1
6	Top Cover		1
7	HX-SHCS 0.138-32x0.5x0.5-S		4
8	HX-SHCS 0.25-20x0.5x0.5-S		12

388. - It is possible to add more columns to the Bill of Materials with information imported from the components' custom properties. To add columns to the table make a right mouse click in the "QTY" cell in the table (or any column you wish), and from the pop up menu select "**Insert, Column Right**" (or Left...) to locate the new column. In the "Column type" select "CUSTOM PROPERTY", and from the "Property name" drop down list select "Material". The value will be automatically filled for components with a "Material" custom property.

389. - Add another column with the custom property "Weight" to complete the table. Open the parts with missing custom properties and add them.

 To open a part's file from within the assembly drawing, select the part in the graphics area with the left mouse button and click in the "Open Part" icon from the pop-up menu.

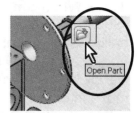

390. - To move the table click and drag it from the upper left corner. Column width and row height can be adjusted as needed and cells, rows, columns or the entire table can be formatted using the pop up toolbar just like Microsoft Excel. When finished your table should look like this.

ITEM NO.	PART NUMBER	DESCRIPTION	QTY.	Material	Weight
1	Housing	Casting No. ABC123	1	Cast Alloy Steel	4.382
2	Side Cover	Part No. SR-123	2	AISI 1020	0.363
3	Worm Gear Shaft	Shaft with Hex Drive	1	Chrome Stainless Steel	0.524
4	Worm Gear	Gear Part No. RG-56789	1	AISI 1020	0.378
5	Offset Shaft	Part No. RT-12345	1	Chrome Stainless Steel	0.466
6	Top Cover	PN. TP-CVR	1	Cast Alloy Steel	0.481
7	HX-SHCS 0.138-32x0.5x0.5-S		4		
8	HX-SHCS 0.25-20x0.5x0.5-S		12		

 We can add as many custom properties as needed to accurately document our designs, and as we learned, these properties can be used in the part's detail drawing and the assembly bill of materials. Assembly files can also be given custom properties.

391. - Now we need to add identification balloons to our assembly drawing to correctly match each item in the assembly view to the item number in the Bill of Materials. Make a right mouse click in the assembly drawing view, and from the pop up menu select "Annotations, **AutoBalloon**" or go to the menu "**Insert, Annotations, AutoBalloon**" after selecting the assembly view.

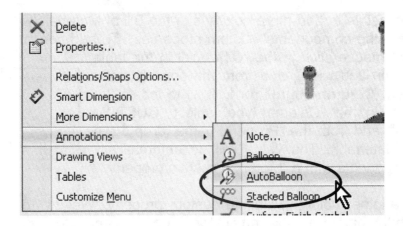

392. - When the AutoBalloon Property Manager is displayed select the options "Square" for the "Balloon Layout", "Ignore multiple instances" to avoid adding balloons to repeated components, "Circular", "2 Characters" and "Item Number" from the "Balloon Settings" options box for the size and style of the balloons. If we click and drag any of the balloons before we click OK, all the balloons will move in or out at the same time. This can help us locate them closer to the view. When satisfied with the general look of the balloons, click OK to add them the drawing view. We can modify their positions as needed by dragging them individually later.

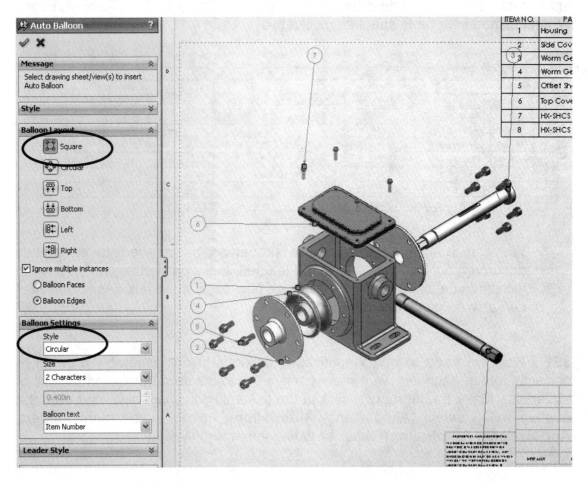

393. - Click and drag the balloons individually to arrange them in the drawing. A balloon's arrow tip can also be dragged to a different area of the part for visibility; if the arrow tip is dragged to a different component the item number displayed will change to reflect the item number of the part that the balloon is attached to. Arrange the balloons as needed for readability.

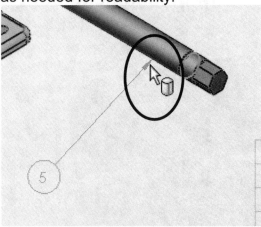

394. - Edit the Sheet Format to add the missing notes in the title block, save the drawing and close the file.

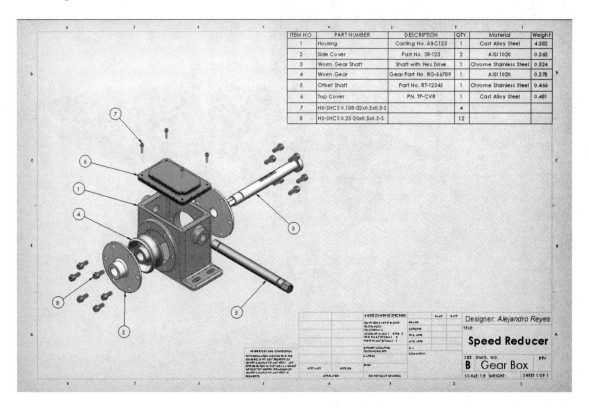

Challenge Exercise:

Make the complete assembly of the Gas Grill using the parts created in the part modeling lesson.

www.mechanicad.com

Analysis: SimulationXpress

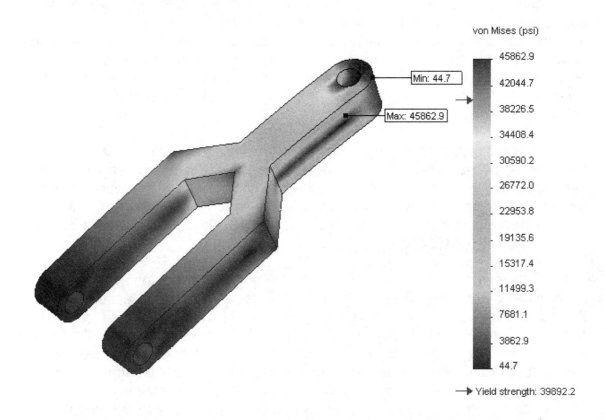

Why Analysis?

Simulation is a very important tool in engineering. Computers and software have come a long way since the early analysis tools became available, and modern tools have made simulation a lot easier to use, faster, accurate and more accessible than ever enabling designers and engineers to check their designs, make sure it's safe, understand how it will deform, if it will fail and under what circumstances or how it will perform in a given environment (temperature, pressure, vibrations, etc.).

The biggest advantage, by far, when analyzing a design is that we will save money by making decisions as to what materials to use, sizes, features and even appearance early in the design process, when it is still all in "paper" (maybe a more appropriate term now would be 'Bites' ☺) and cheaper to modify. As the product development process advances making changes to a product's design are increasingly more expensive.

Picture it this way: Imagine we design a new cellular phone that looks really nice. We make molds, tooling, order parts, and setup an assembly line. Soon the phones start selling, and a month later, we start getting customer complains; when the call button is pressed hard the battery cover falls off. Then, *after* we know we have a problem, we run an analysis and find out that *we should have* had a thicker *this or that*, with a *widget* in between the *thingamajig* and the *thingamabob*, and those changes would solve the problem. Now, all we need to do is to change the design, the tooling, the assembly line, marketing, and above all, convince every customer that it is fixed. At this point, our customer's perception of our company and credibility are destroyed. This is usually the most expensive. It's easy to see why making analysis of our products early in the process will help us design better products and save resources and money.

With that said, it also has to be noted that the simple fact of making a simulation does not guarantee that our designs will be successful, as there are many factors involved. Reasons for product failure include using the product beyond its designed capacity, abuse, material imperfections, fabrication processes and things and circumstances we could have never thought of. This is where the designer needs to take into account every possible scenario, and of course, simulate it as realistically as possible with the correct analysis tools.

SolidWorks includes a basic analysis package called "**SimulationXpress**". As its name implies, it has limited functionality that allows only certain scenarios to be analyzed. In order to understand what these limitations are, first we need to learn a little about how analysis works. A general overview of how it actually works is as follows:

Analysis, also known as "Finite Element Analysis", is a mathematical method where we divide a component's geometry into hundreds or thousands of small pieces (elements), where they are all connected to one or more

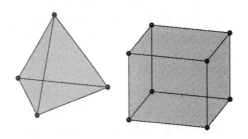

neighboring elements by the vertices (called nodes in analysis). The elements have simple geometry that can be easily analyzed using stress analysis formulas. Elements are usually tetrahedral or hexahedral for most solid models. There are other types of elements including shell and linear elements. SimulationXpress uses tetrahedral elements.

In order for us to make a simulation we need to know what the component is made of (physical properties of the material), how it is supported (Restraints), and what forces are acting on it (Loads). The next step is to break down the model into small elements, this process is called meshing. The result is called the **Mesh**; in SimulationXpress the only input we give, if we choose to, is to define the average element size. SimulationXpress automatically creates the mesh for us.

The next step is to define how the component is supported, in other words which faces, edges or vertices are restrained. Each node has six degrees of freedom, meaning that each node can move along X, Y and Z and rotate about the X, Y and Z axes. Restraints limit the degrees of freedom (DOF) of the nodes in the face or edge that they are applied to. SimulationXpress limits the user to restrain (Fix) all DOF in faces, meaning that they are completely immovable.

This approach is appropriate for simple analyses, but in order to obtain more accurate solutions a more realistic approach should be used.

Now that we know how the model is supported, we need to define the Loads (forces) that act upon it. SimulationXpress only allows Forces and Pressures to be applied to model faces. We can define the loads to be normal to the face they are applied to or perpendicular to a reference plane.

SimulationXpress is limited to **Linear Static Stress Analysis**. Stress is defined as the internal resistance of a body when it is deformed and is measured in units of Force divided by Area. For example, if we load a bar with 1 in^2 cross section area with a 1 Lb load, we'll have a stress of 1 Lb/in^2. The stress depends on the forces and the geometry regardless of the material used. The material properties will make a difference as the model will deform more or less.

The "Static" part means the model is immovable; in other words, the restraints will not let the model move with the applied loads. Loads and restraints are in equilibrium.

The "Linear" part implies the deformation of the model is proportional to the forces applied, twice the force, twice the deformation. Deformations are generally small compared to the overall size of the part and occur in the **Elastic** region of the "Stress-Strain" curve of the material used. If we remove the force, the model returns to its original shape. It's like a spring: If we apply a force it will deform, twice the force, twice the deformation. If we remove the force, the spring returns to the original size.

In general terms, the **Yield Stress** is the point where the stress-strain curve is no longer lineal. If a material is stressed beyond the Yield Stress it will be permanently deformed. In this case we will have **Plastic Deformation**. Thinking of the spring before, if we pull the spring too far it is permanently deformed, meaning it had plastic deformation. Once the yield stress is exceeded the analysis results of SimulationXpress are invalid. In this case we need to

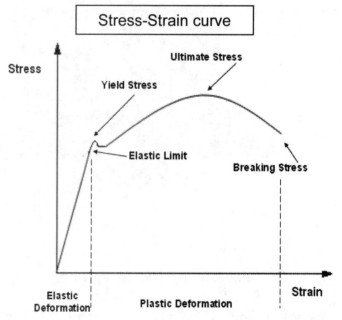

change our design, geometry, loads or material so that the maximum stress in the model does not exceed the material's yield stress.

395. - To show the SimulationXpress functionality, complete the following part to analyze it. Assign the material "Aluminum 6061-T6". Tip: Make the part with square ends and round them using a "Full Round Fillet".

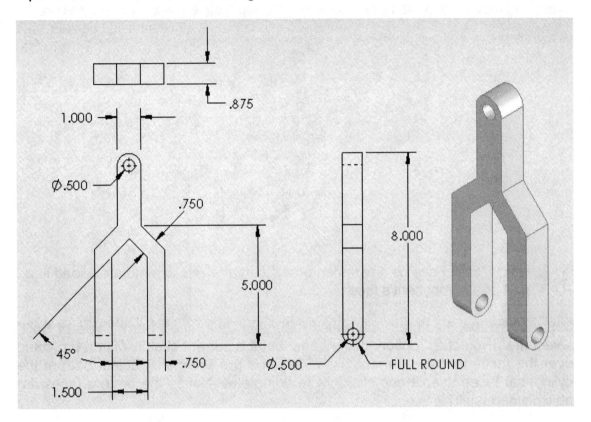

396. - We will add forces in the bottom holes perpendicular to the model and support it from the top hole.

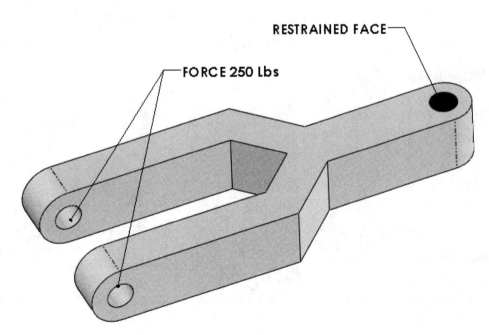

397. - In order to better simulate the effect of the force, we need to divide the bottom faces in two, as the force will only be acting on one half of the cylindrical face. To accomplish this we will use the "**Split Line**" command. First make a sketch as indicated in the right side of the model. Notice it is only a single line.

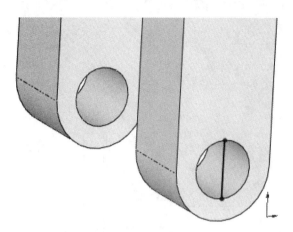

 Using "Split Face" is a common practice in analysis to simulate a load in a part of a component's face.

398. - From the "Features" tab, select "Curves, Split Line"; from "Type of Split" select the "Projection" option; this means that the current sketch will be projected over the faces to split. "Current Sketch" will be pre-selected. Select both of the cylindrical faces to split and click OK to complete. Notice the bottom faces are now divided (split) in two.

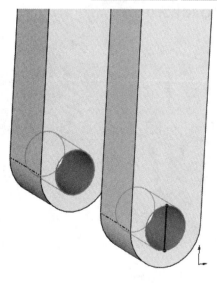

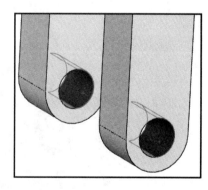

399. - Now that we have prepared our model for analysis, activate the SimulationXpress wizard using the menu "**Tools, SimulationXpress**" or in the "Evaluate" tab in the Command Manager.

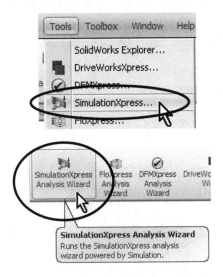

400. - Starting with the SolidWorks 2010 version the SimulationXpress wizard is now integrated in the Task Pane. Select "Options" to set the units to PSI and turn on the checkbox to display the annotations for minimum and maximum in the result plots. Click Next to continue.

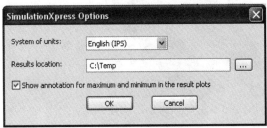

401. - The first step is to define the "Fixtures", referring to the faces that will be completely rigid. We are also given a warning letting us know that the results in the regions close to the fixtures may be unrealistic, and more accurate results can be achieved with *SolidWorks Simulation Professional* by simulating a more realistic condition. After selecting "Add a Fixture" we are presented with the Fixture selection box; select the face at the top and click OK. For this example we only need one fixture.

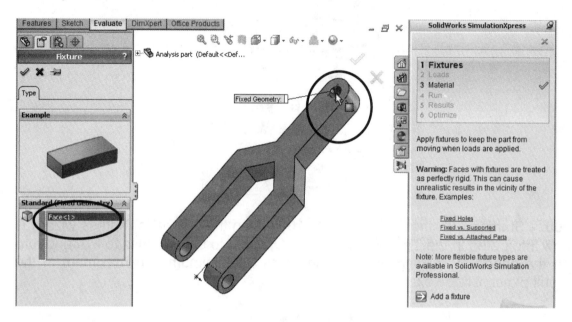

402. - If more Fixtures are needed select "Add a Fixture" as needed. The SimulationXpress Study now shows the "Fixed-1" condition and "Fixtures" shows a green checkmark letting us know this step is complete. Click Next to continue.

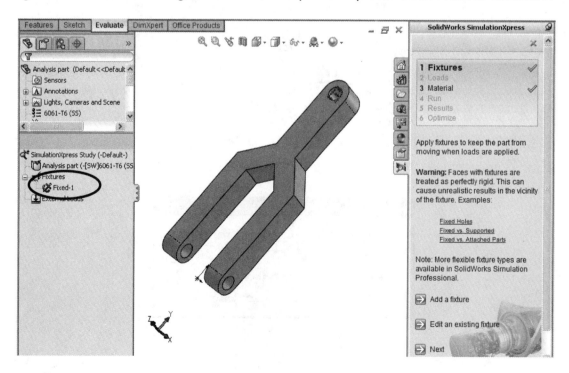

403. - Now we need to apply the forces or pressures that will act on the part. The forces that we are adding will be 250 lbf perpendicular to the part's Front plane in each of the two bottom holes. Select "Add a Force" to show the "Force" dialog box, here we'll select one side of each of the two faces we split earlier; make the force value 250 lbf. To make the forces perpendicular to the part, select the option "Selected Direction" and from the Fly-out-Feature Manager select the "Front" plane. Make sure the option "Per Item" is selected, as we want to apply the same 250 lbf to each face, otherwise the 250 lbf force will be equally distributed over the faces selected. If needed turn on the "Reverse Direction" box. In this example it doesn't make any difference since the part is symmetrical. Click OK when done and select "Next" from the SimulationXpress wizard at the right.

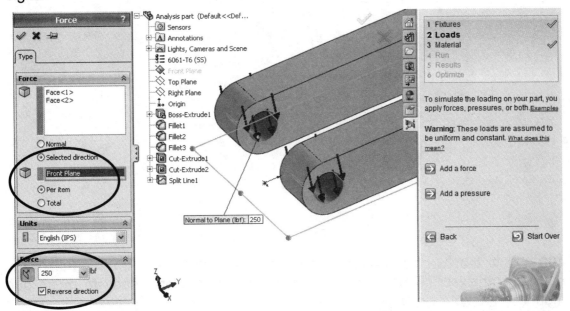

404. - The "Force-1" condition is added to the SimulationXpress Study and we can see the preview of the forces applied to the faces and their direction. Now "Loads" shows a green checkmark letting us know this step is complete. If we need to add more forces or pressures to other model faces, or modify this one, we would do it in this step. Click Next to continue.

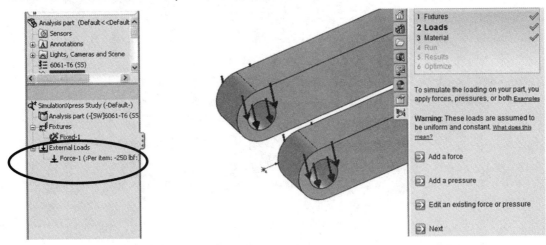

405. - The next step is to define the material that will be used for the part. This is important because as we explained earlier we need to know the physical properties of the material being analyzed. Since we had assigned a material before the analysis (Aluminum 6061-T6), this step is already complete. We can see the green checkbox next to "Material" in the SimulationXpress wizard as well as the Young's Modulus and Yield Strength values for reference. If we had forgotten to define a material or the units are not listed using the units we are used to, we can select "Change Material" and change it from the materials list. Click "Next" to continue.

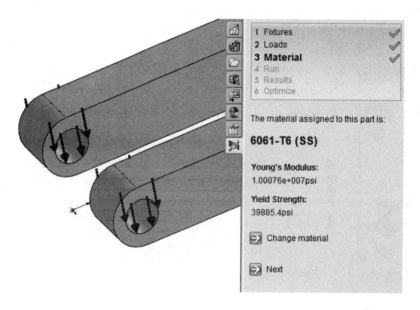

406. - Now that we have given SimulationXpress the minimum information to make an analysis (Fixtures, Forces and Material), we are ready to mesh our model and run our simulation.

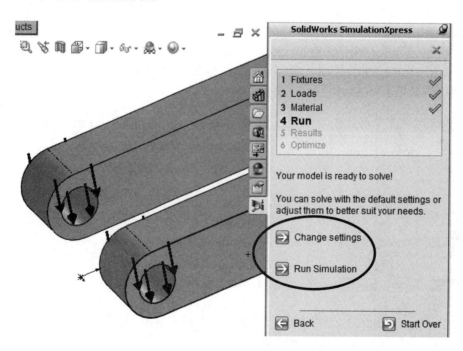

407. - At this point we have an option to run the analysis with the default settings for the mesh or change them. Using a smaller mesh size will generate more elements and will provide a more accurate solution; however, increasing the number of elements also increases the time and computing resources needed to run the simulation. As a general rule, if our maximum stresses are close to the acceptable limits try using a slightly smaller mesh element. SimulationXpress allows us to change the mesh density, or the general size of the elements under "Mesh Parameters". From the wizard select "Change settings" and then "Change mesh density". Click OK to use the default mesh settings.

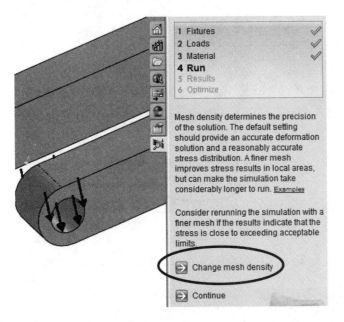
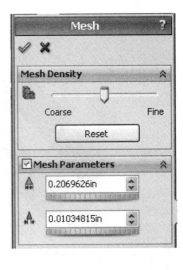

408. - As soon as we click OK in the "Mesh" dialog box, SimulationXpress starts meshing the model, in other words, breaking the model into many small pieces (*Finite Elements*). Then we can see the meshed model ready for analysis.

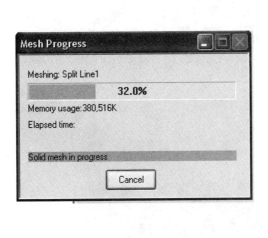

409. - Now that the model is meshed select "Run Simulation" from the SimulationXpress wizard. What happens next, in the simplest terms, is that the solver will generate an equation for each element, with an unknown variable at every node. Since adjacent elements share nodes between them we end up with a very large matrix of hundreds or thousands of simultaneous equations. The solver calculates all the unknown variables at each node and the result for each node is the node's displacement or deformation; based on the physical properties of the material we can calculate the stresses in the model at each node.

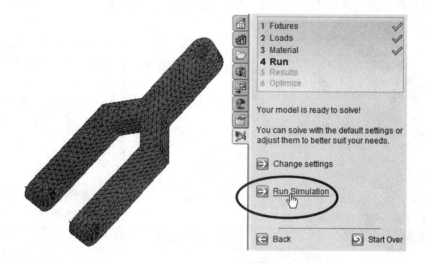

Since this is a small model it only takes a few seconds to finish the analysis.

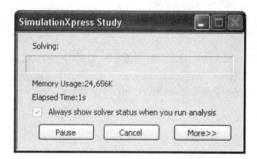

410. - When the solver is finished we see an animation of the deformed model.

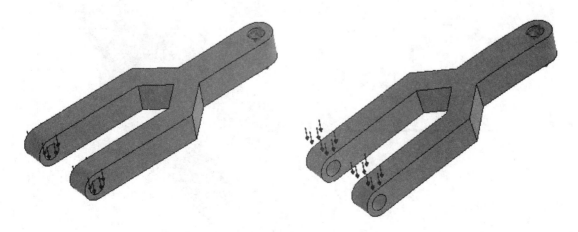

411. - The model deforms as we expected in the direction of the forces, so we can proceed to view the results. If the model had not deformed as expected we could go back to modify the fixtures and loads. Select "Yes, continue" for the results.

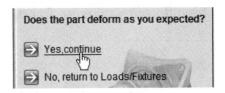

The first results we see are the stresses calculated using the **von Mises** method which is most commonly used for isotropic materials (materials with the same physical properties in every direction), mostly metals. The deformation is exaggerated for visual clarity (In this case 7.53 times).

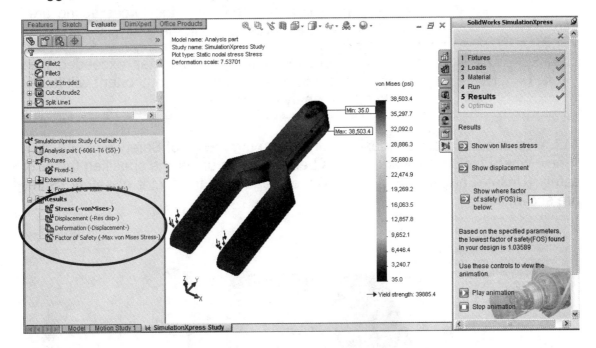

412. - In these results we can see the maximum stress in the model is approximately 38,500 psi, very close to the yield strength of 39,885 psi. Based on these results the lowest Factor of Safety in our model is 1.04. The color scale will serve as a guide to finding the stresses in the rest of the model.

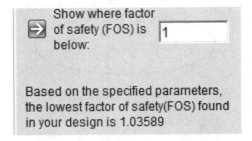

The factor of safety is calculated by dividing the material's yield strength by the stress at each node. We are interested in finding the lowest value because if any area of the model has a factor of safety below 1 this means the stress is higher than the Yield Strength and our model will have permanent deformation and will be considered to fail.

 Note that we are talking about permanent deformation and not breaking; to yield not necessarily means that it will break; it will break if we reach the Breaking Stress. Permanent deformation is also considered to fail.

The different values for **Factor of Safety** (FOS) mean:

Value		Meaning
> 1	✓	The stress at this location in the part is less than the yield strength and is safe. Depending on the specific application we may have to design with a higher FOS to ensure our design is safe and have room for unexpected loads.
= 1	✗	The stress at this location in the part is exactly the yield strength. This part has started or is about to yield (deform plastically) and will most likely fail if the forces acting are slightly more than the forces used in the analysis.
< 1	✗	The stress at this location in the part has exceeded the yield strength of the material and the component will fail if subjected to the analysis parameters.

413. - To see how safe our design is, type 2 in the Factor of Safety box and click "Show where factor of safety (FOS) is below". Areas under the specified factor of safety (Unsafe) are shown in red and the areas over the factor of safety (Safe) are shown blue.

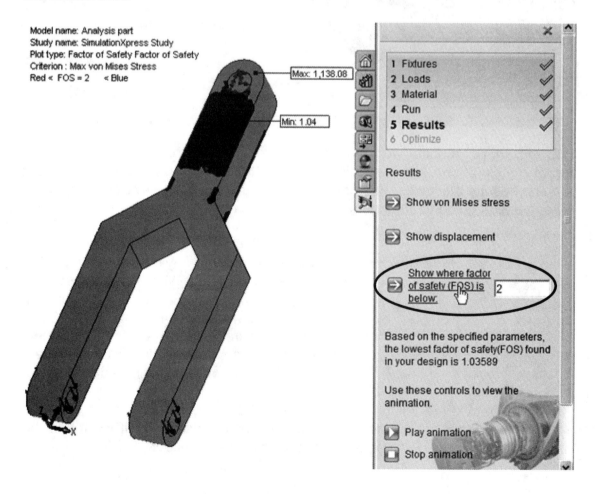

414. - Activate the "**Displacement Results**". This way we can see how much our model deforms under the current conditions. The maximum deformation is 0.1" at the bottom of the part, as expected since the top is rigidly supported (The values in the scale are listed in scientific notation).

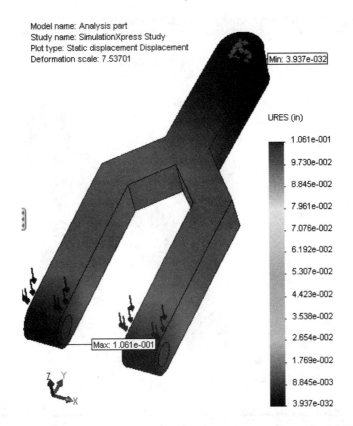

415. - Select "Done viewing results" to move to the next screen where we can generate reports either in HTML or eDrawings format. Selecting "Generate HTML report" will  generate the files needed to view the report in a web browser. The report will include the following information as well as screen captures of results and details of the analysis including loads, fixtures and material properties using the current model orientation.

1. **File Information**
2. **Materials**
3. **Load & Restraint Information**
4. **Study Property**
5. **Results**
    a. **Stress**
    b. **Displacement**
    c. **Deformation**
    d. **Factor of Safety**
6. **Appendix**

416. - Selecting "Generate eDrawings file" will make an eDrawings file with the analysis results. The main difference from the HTML report is that in eDrawings we can rotate the models, turn the mesh on or off, zoom in or out and easily share the results with other people.

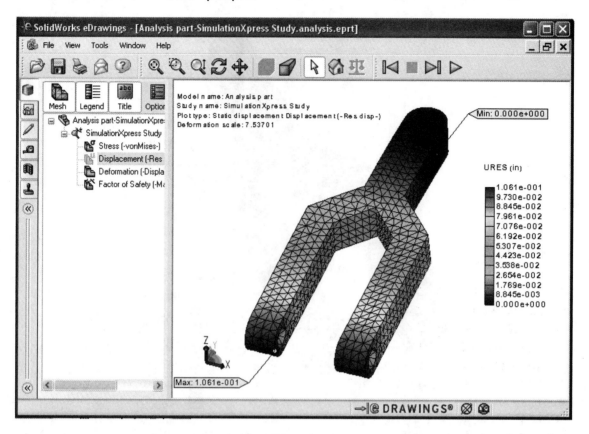

417. - What we are going to do now is to change a dimension in the model and re-run the analysis to see how much the results change. Click to close the SimulationXpress wizard. When asked if we want to save the results select "Yes" from the dialog box.

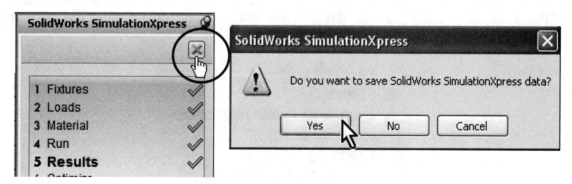

418. - For this example change the 0.875" thickness dimension to 1.125" and rebuild the model.

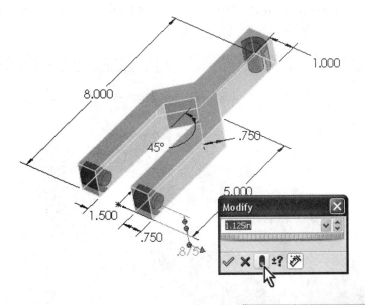

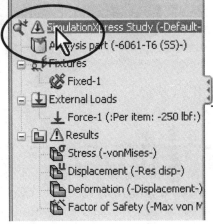

419. - After rebuilding the model select the SimulationXpress wizard icon to go back to the analysis. Notice the warning icons in the SimulationXpress study. This warning icon indicates that we have a problem, in this case the results are no longer valid because we changed the model, and since the geometry is different the mesh is different and therefore the results are not valid.

420. - To update the results all we need to do is to make a right mouse click at the top of the analysis tree and select "Run" from the pop-up menu to update the results. SimulationXpress will automatically re-mesh the model and run the analysis.

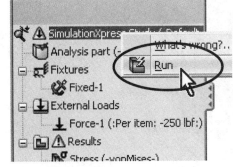

421. - After running the analysis with the updated geometry we see the updated results, and now we have a maximum von Mises stress of 24,837 psi, giving us a minimum factor of safety of 1.6 and a maximum displacement of 0.05".

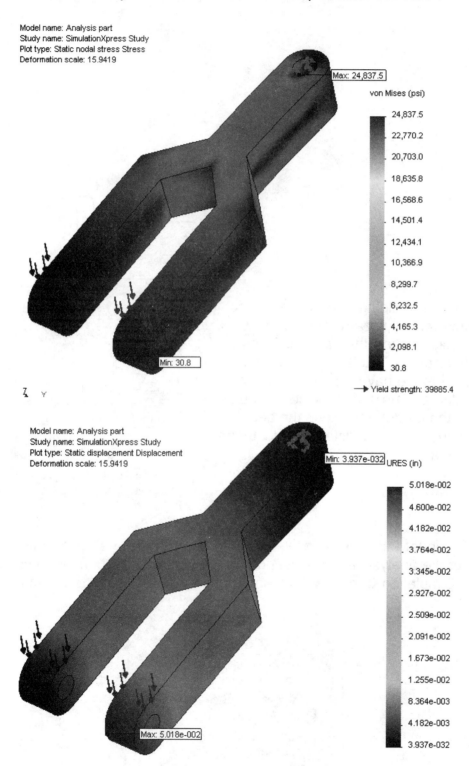

422. - CHALLENGE: SimulationXpress implemented a new interface for the Optimization wizard that can be accessed after reviewing the results and generating the reports. Select the option to optimize your model. The optimization wizard will help us find the lightest model (minimum mass) that complies with the constraints defined. SimulationXpress allows only one dimension to be varied within a range of values defined by the user; the constraints can be a minimum factor of safety, a maximum von Mises stress or displacement. Run the analysis and try to find a better solution that meets a minimum factor of safety of 1.5 by changing the same dimension we modified.

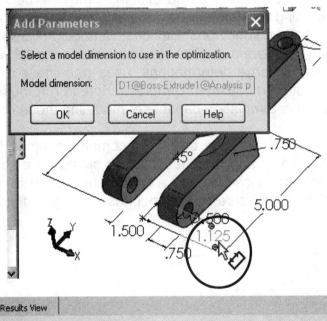

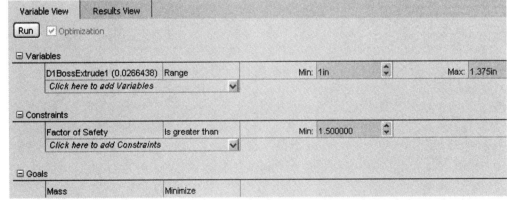

 A note on optimization: in order for the optimization to work we need to have a study with better results than what we need, basically *over designed*. For example, if the study has a high factor of safety, we can optimize the study to make the model lighter, and still have a factor of safety high enough to be a safe design.

A final word on analysis: It is very important to make the correct assumptions for restrains and loads applied to a model, as the accuracy and reliability of the results depend directly on them. Talking about results, it's just as important to understand what the results of the simulation mean in order to make the correct deductions, and not over design, or put our designs at risk of failure.

As we explained before, SimulationXpress is a *first-pass-analysis* tool. This means that its purpose is to give us a general idea of how the design will perform with the conditions given. If our design is not *comfortably* safe, and/or we know this is a critical component of the design, it may be best to make a more complete analysis using the full suite of SolidWorks Simulation software where we can add more realistic conditions. Besides, SimulationXpress is only capable of linear static analysis, therefore, if we anticipate that a model will be exposed to temperature, vibrations, large deformations, uses non-linear materials, or other conditions that cannot be modeled in SimulationXpress run those analyses using a version of SolidWorks Simulation capable of modeling those conditions.

Analysis and simulation can be a very complex subject depending on the physical phenomena being analyzed. It can be as simple as a stress analysis in a link like this example or as complex as a space ship taking into account forces, temperature, vibrations, pressure, radiation, phase changes, fluids flow, etc.

Collaboration: eDrawings

How to collaborate with non-SolidWorks users

Sometimes we need to collaborate with other members of the design team in the same room, same building, or even across the world. With the convenience of Email, it's easier than ever to share our designs using eDrawings; we can send our extended design team, customers or suppliers a file by email with all the design information needed for them to review and send feedback to the designer.

Starting with SolidWorks 2010 eDrawings is completely embedded in SolidWorks and is no longer separate software that needs to be loaded and is always available.

We can generate eDrawings files from Parts, Assemblies, Drawings and simulation studies. The only thing we need to do is to press the Publish eDrawings File icon from the save fly-out toolbar or go to the menu "**File, Publish eDrawings File**" while the document we want to share is active. The eDrawings viewer will be loaded with the same document. After we generate an eDrawings file it becomes a read only file, we cannot modify it but we can print it, zoom in or out, section it, etc. For this example we'll publish an eDrawings file from the assembly drawing. eDrawings will look like this:

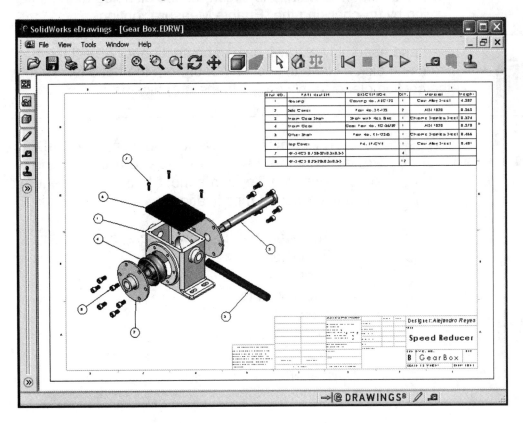

If the eDrawings file is generated using SolidWorks Office Professional or the educational edition, the eDrawings Professional version is loaded. eDrawings Professional includes additional options, including tools to move the components in assemblies, measure, add markups and stamps, and if it's a part or assembly, make section views of the models.

The Previous, Stop, Next and Play buttons will allow you to navigate different views if you have a multiple view drawing, or to go from one view orientation to the next in Parts and Assemblies. Pressing the play icon will animate the views, and will change from one to the next. At any time, the animation can be stopped, and the view zoomed in, rotated and panned.

	While in an eDrawings of an assembly, the parts can be moved by dragging them. To return them to their original location, double click on them or click in the Home icon.
	If the Assembly has an exploded view, the Explode button can be used to explode or collapse the assembly.
	Part, Assembly and Drawing eDrawings can be measured. To help protect confidential data, this option can be disabled at the time of saving by un-checking the "Enable Measure" option.
	Part and Assembly eDrawings can be sectioned. Select the Section icon, and pick a plane to section about from the options. To move the section plane simply drag it in the graphics area.

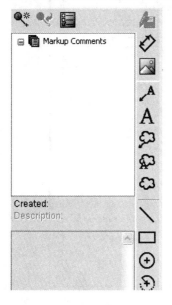

The most powerful option of eDrawings Professional are the markup tools to allow anyone to add annotations to a file, and send it back to the generator to make the required modifications.

The markup tools allow us to add:
- Dimensions
- Notes
- Clouds with text
- Lines
- Rectangles
- Circles
- Arcs and
- Splines, etc.

After creating the eDrawings file, we have to save it. eDrawings allows us to save the files different ways.

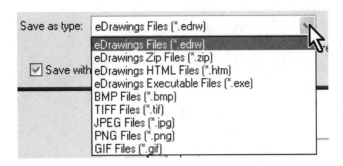

Saving as eDrawings Files (*.edrw) will result in the smallest file size, however, the recipient needs to have the free eDrawings viewer loaded in their computer.

Saving as an Executable file (*.exe) will add the viewer to the file in a single file. The down side of this is that emailing an Executable file may be rejected by the recipient's email. To work around this problem, the file may be saved as a Zip file. This will be an Executable file zipped for email. To view these files, the recipient does not need to have the eDrawings viewer installed.

Saving as an HTML File (*.html) will create a web page that can be emailed. eDrawings will be loaded automatically if it is already installed in the computer, if it is not installed, the web browser will ask to install and run the corresponding plug-in. Optionally, a link to download the eDrawings viewer is listed at the bottom of the web page.

Notes:

Final Comments

By completing the exercises presented in this book, we learned how to apply many different SolidWorks features and its various options to common design tasks. As we stated at the beginning, this book is meant to be an introduction to SolidWorks, and as the reader was able to see, the breadth of options and possibilities available to the user are enough for any design task at hand. After completing this book the reader is capable of accomplishing many different tasks from part and assembly modeling to detailing in a very short period of time, using the most commonly used commands, as well as making a stress analysis with the included SimulationXpress software.

We hope this book serves as a stepping stone for the reader to learn more. A curious reader will be able to venture into more advanced features, and take advantage of the similarities and consistency of the user interface to his/her advantage. We tried very hard to make the content as understandable and easy to follow as possible, as well as to get the reader working on SolidWorks almost immediately, maximizing the "hands-on" time.

After completing all parts, drawings, an assembly including the fasteners, exploded view and the assembly drawing complete with a Bill of Materials, the reader is ready to apply the learned concepts in different design applications.

One final tip: if you get lost, or can't find what you are looking for while working in SolidWorks, click with the Right Mouse Button. Chances are, what you are looking for is in that pop up menu. ☺

Notes:

Appendix

Document templates

One of the main reasons to have multiple templates is to have different settings, especially units; we can have millimeter and inch templates, different materials for part templates, dimensioning standards, etc. and every time we make a new part based on that template, the new document will have the same settings of the template. One good idea is to have a folder to store our templates, and add it to the SolidWorks list of templates.

Using Windows Explorer, make a new folder to store our new templates. For this example we'll make a new folder in the Desktop called *"MySWTemplates"*. In SolidWorks, go to the menu "**Tools, Options**" and in the "System Options" tab, select "File Locations". From the drop down menu select "Document Templates".

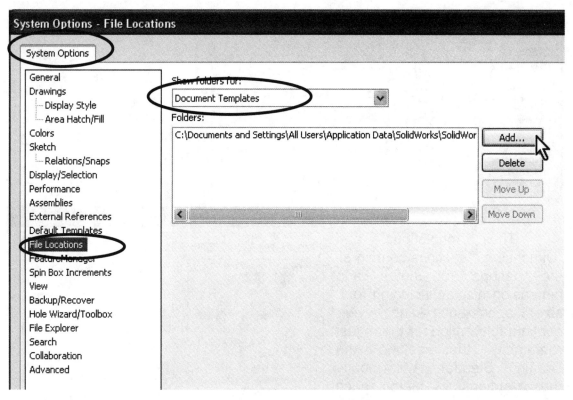

Click the "Add" button and locate the folder we just made to save the templates. Click OK when done. You will now see two template folders listed.

To create a new template make a Part, Assembly or Drawing file, and go to the menu "**Tools, Options**" and select the "Document Properties" tab. Everything we change in this tab will be saved with the template. The most common options changed in a template are Units, Grid, detailing standards and Annotation Fonts. Let's review them in the order that they appear in the options.

The first section is "Drafting Standard"; here we can change the Dimensioning standard to ANSI, ISO, DIN, JIS, etc.; by changing the standard, all the necessary changes will be made to arrows, dimensions, annotations, etc. according to that standard.

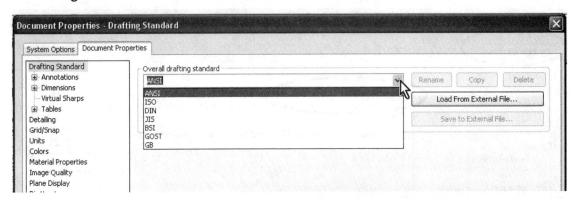

The "Annotations" and "Dimensions" sections contain all the options for notes and dimension display including font, size, appearance and every configurable option for them; just remember that setting the Drafting Standard will modify these options to match each standard.

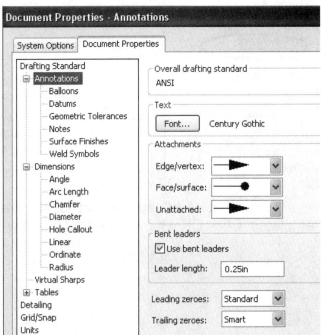

In the "Detailing" area we can change what is displayed in parts and assemblies; for drawings we have more options including which annotations are automatically inserted when a new drawing view is created.

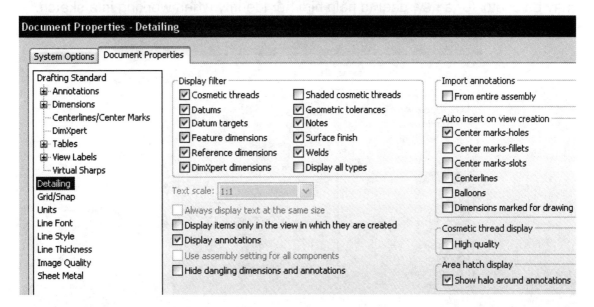

In the "Units" section select the units that we want to use in the template, number of decimal places or fractional values, angular units, length, mass, volume and force. Selecting a unit system will change all the corresponding units to that system, saving us time. Selecting "Custom" will allow us to mix unit systems if so desired.

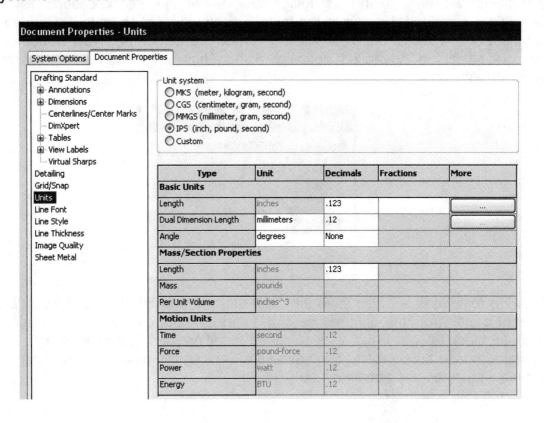

For the "Grid/Snap" section, we can define if we want to have a grid and its settings when we work in a sketch. Most users don't use the grid, but that's precisely why it's called an option. Turn it on if you like it; turn it off if you don't. It may be useful for a new user to help him/her identify when working in a sketch.

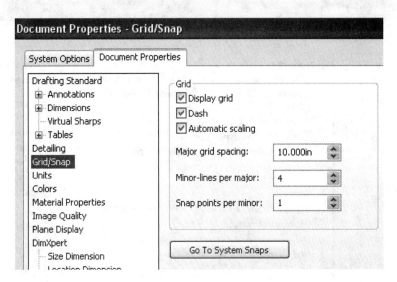

Click OK to change the document properties to the options selected. To save the documents as a new template go to the menu "**File, Save As**". From the "Save As Type" drop down box, select "Part Template (*.prtdot)", "Assembly Template (*.asmdot)" or "Drawing Template (*.drwdot)" (depending on the type of document we are working on you will only see one option). SolidWorks will automatically change to the first folder listed under the "File Locations" list for templates; if needed, browse to the folder where we want to save it. Give your template a new name and click "Save". When we select the "New Document" icon and select the "Advanced" option we'll see a new templates tab with our new template listed in it.

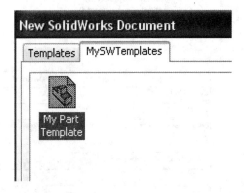

Drawing templates will also save the Sheet format and title block. Select "Edit Sheet Format" from the right mouse click pop up menu, change the annotations and title block to your liking; when finished modifying it select "Edit Sheet" from the right mouse button menu and save as a drawing template. When using this template all notes and annotations will be set.

Index